Ilaria Damiani

Untwisted Affine Quantum Algebras: the Highest Coefficient of $\det H_\eta$ and the Center at Odd Roots of 1

TESI DI PERFEZIONAMENTO

SCUOLA NORMALE SUPERIORE
PISA - 1996

Tesi di perfezionamento in Matematica sostenuta il 18 ottobre 1995

COMMISSIONE GIUDICATRICE

Edoardo Vesentini, Presidente
Corrado de Concini
Roberto Dvornicich
Claudio Procesi
Marc Rosso
Giuseppe Tomassini
Carlo Traverso
Mario Salvetti

ISBN: 978-88-7642-285-0

INDEX

INTRODUCTION.

In this thesis I study the untwisted affine quantum algebras $\mathcal{U}_q = \mathcal{U}_q(\hat{\mathfrak{g}})$ (here $\mathfrak{g}$ is a simple finite dimensional Lie algebra and $\hat{\mathfrak{g}}$ is a central extension of its loop algebra) and their specialization at ε, where ε is a primitive l^{th} root of unity, with l bigger than 1 if the Dynkin diagram is simply laced, bigger than 3 if $\mathfrak{g} = G_2$ and bigger than 2 otherwise (see remarks 1.D1.15 and 1.D1.16 for a discussion about the conditions imposed on l). In particular my goal is the complete description of the center $\mathcal{Z}(\mathcal{U}_\varepsilon)$, when l satisfies the additional condition of being odd.
The center of the specialization of the quantum algebra at odd roots of unity is already known in the finite case, that is, for finite dimensional semisimple Lie algebras (see De Concini Kac [6]), and in this thesis I prove that analogous results hold in the affine untwisted case.
The main tool is the study of the contravariant form H defined on the highest weight modules of $\mathcal{U}_q$, and in particular on the Verma modules (see §1.D4). Indeed the investigation of the connections between H and the center $\mathcal{Z}(\mathcal{U}_\varepsilon)$ shows that $\forall \eta \in Q_+$ the multiplicity of ε in $\det H_\eta$ gives an upper bound for the dimension of $\mathcal{Z}(\mathcal{U}_\varepsilon) \cap \mathcal{U}^+_{\varepsilon,\eta}$: see lemma 3.2.10.
Then my program is (i) the calculation of the multiplicity of ε in $\det H_\eta$, and (ii) the exhibition of enough central vectors.
For (i) I first find the highest coefficient b_η of the determinant $\det H_\eta$, which is the coefficient of $K_{\mathrm{par}(\eta)\eta}$ in $\det H_\eta$ over a suitable integer form of $\mathcal{U}_q$ and is well defined up to units of $\mathbb{C}[q, q^{-1}]$ (proposition 2.1.3): to this aim I use a PBW basis of $\mathcal{U}_q$ (see Beck [2]) and some properties of H, which allow to reduce the problem to the calculation of the determinant of a matrix which can be transformed into a triangular matrix (see §2.2). The real root vectors behave "well", in a sense which will be clear (corollary 2.3.4), that is, the "real part" of the matrix whose determinant is b_η is already triangular. For the imaginary root vectors a little more work is necessary in order to give the matrix a triangular form: this will be done in §2.5. and §2.6.
I shall then use a "Casimir" operator C, which is defined on the highest weight modules of $\mathcal{U}_q$ and gives conditions on a vector to be primitive (see Tanisaki [23]), to prove that b_η divides $\det H_\eta$ over

$$\mathcal{A} \doteqdot \mathbb{C}[q, q^{-1}, (q^r - q^{-r})^{-1} | 1 \leq r \leq \max\{d_i | i \in I_0\}]$$

(see proposition 2.1.10); thus the multiplicity of ε in $\det H_\eta$ is exactly the multiplicity of ε in b_η, and knowing b_η is enough to compute the upper bound for the dimension of $\mathcal{Z}(\mathcal{U}_\varepsilon)$ that I am looking for.
I then devote my effort to looking for enough central elements in $\mathcal{U}^+_\varepsilon$: some of them are easily constructed as suitable powers of real root vectors (corollary 3.3.3), exactly as in the finite case, but the imaginary case presents some more complications: there are natural imaginary root vectors which are central in $\mathcal{U}_\varepsilon$ (proposition 3.4.4), but while "in general" these vectors are enough to describe the whole center, there are special values of l when some degeneracies

occur. In these cases the triangularization of the matrix $\det H_\eta^{\max}$ provides in a natural way the central vectors which are still missing (see propositions 3.4.6 and 3.4.8).
Comparing the total number of central elements thus exhibited with the multiplicity of ε in $\det H_\eta$ one finds $\mathcal{Z}(\mathcal{U}_\varepsilon) \cap \mathcal{U}_\varepsilon^+$.
It is then easy to describe $\mathcal{Z}(\mathcal{U}_\varepsilon)$ completely, using the particular properties of stability with respect to the action of the Weyl group holding for the imaginary root vectors.
The final result is summarized in theorem 3.5.6.
The first chapter is essentially a dictionary of the main notions, recalling the fundamental results about finite and affine root systems and untwisted affine quantum algebras, with particular stress on the construction of PBW-bases, Verma modules and specializations.

CHAPTER 1. PRELIMINARIES.

In this chapter I shall give a description of the general setting in which the theory of quantum affine algebras developes, and state most results that I will need in the following.

In Part A. I define the Kac-Moody algebra $\mathfrak{g}$ associated to a generalized Cartan matrix A, and the related Dynkin diagram, concentrating on the structure of the root system and on the action of the Weyl group. After studying the finite case, I restrict my attention to the untwisted affine case, describing the root system, the Weyl group and the extended Weyl group, in relation with those of a suitable finite Cartan matrix. I recall in this part the classification of finite and untwisted affine Dynkin diagrams, stressing the strict relation between the two families of diagrams.

In Part B. I define the quantum algebra $\mathcal{U}_q$ associated to a symmetrizable generalized Cartan matrix, and study its structure, in particular the triangular decomposition and the Q-gradation. I also introduce the braid group $\mathcal{B}$ and its extension $\tilde{\mathcal{B}}$ by the group $\mathcal{T}$ of automorphisms of the Dynkin diagram, and study its action on $\mathcal{U}_q$. Finally I recall that $\mathcal{U}_q$ can be provided with a structure of Hopf algebra, and state some of its properties (for example the existence of the adjoint reprensentation), concluding with the assertion of the existence of a pairing between $\mathcal{U}_q^{\geq 0}$ and $\mathcal{U}_q^{\leq 0}$ (see Tanisaki [23]).

In Part C. I recall Beck's method to construct PBW-bases in the untwisted affine case, I study some properties of these bases and show that the hypotheses for Beck's construction to give a PBW-basis of $\mathcal{U}_q$ can be somehow weakened, which implies a useful result about the action of the braid group on $\mathcal{U}_q^+$ and $\mathcal{U}_q^-$. Then I choose one of these PBW-bases and, following Beck, I describe more precisely some relations among the positive root vectors, giving, for each vertex i of the associated finite Dynkin diagram, an injection of $\mathcal{U}_{q_i}(\widehat{sl(2)})$ in $\mathcal{U}_q$, where q_i depends on i.

In Part D. I introduce the notion of integer form of $\mathcal{U}_q$ and the specialization $\mathcal{U}_\varepsilon$ of $\mathcal{U}_q$ at a nonzero complex number ε. In particular, I am interested in the case when ε is a primitive l^{th} root of 1, and I explain in details the restrictions one needs on l so that $\mathcal{U}_\varepsilon$ inherits from $\mathcal{U}_q$ the triangular decomposition, the $\mathbb{C}$-antilinear antiinvolution Ω and the action of the braid group (indeed of the extended braid group): l shall be an integer greater than 1 in the simply laced case, greater than 3 if $A = G_2$, and greater than 2 otherwise; a further restriction will be needed for calculating $\mathcal{Z}(\mathcal{U}_\varepsilon)$, so that I will generally require that l be also odd.

Then I deal with Verma modules. I extend the scalars from $\mathbb{C}(q)$ to an algebraic closure $\Bbbk$ of $\mathbb{C}(q)$; then for each character of $\mathcal{U}_{\Bbbk}^0 \doteqdot \mathcal{U}_q^0 \otimes_{\mathbb{C}(q)} \Bbbk$ (that is, for each $\varphi \in (\Bbbk^*)^I$) one can define a Verma module of highest weight φ. I shortly recall how to define the integer form and the specialization of Verma modules, and describe some of their properties. I define here the contravariant form H, which will play a central role in chapter 2, devoted to the study of the determinant of H over suitable finite dimensional spaces.

I also introduce, following Tanisaki, the "Casimir" operator C; the action of C on each Verma module is diagonalizable and the eigenvalues can be calculated; moreover, C leaves each primitive vector fixed. This allows to give conditions on φ so that primitive vectors exist in the Verma modules of highest weight φ, which will be a fundamental step in the study of H.

Part A. General setting.

§A1. The Kac-Moody algebra; Dynkin diagram, root system, Weyl group.

In this section I introduce the notion of generalized Cartan matrix A and some objects associated with A. For the details see Bourbaki [4] and Kac [12].

Definition 1.
$A = (a_{i,j})_{i,j\in I} \in \mathfrak{M}_{n\times n}(\mathbb{Z})$ is said to be a generalized Cartan matrix if:
1) $a_{i,i} = 2 \ \ \forall i\in I$,
2) $a_{i,j} \leq 0 \ \ \forall i,j\in I$ with $i \neq j$,
3) $a_{i,j}a_{j,i} = 0 \Leftrightarrow a_{i,j} = a_{j,i} = 0$.
A generalized Cartan matrix A is said to be indecomposable if for every nonempty proper subset J of I $\exists i\in J$, $j \notin J$ such that $a_{i,j} \neq 0$.
A is said to be symmetrizable if $\exists D = \mathrm{diag}(d_i|i \in I) \in \mathfrak{M}_{n\times n}(\mathbb{Z})$ ($d_i > 0 \ \ \forall i\in I$) such that DA is symmetric (the d_i's are always chosen such that g.c.d.$(d_i|i\in I)$=1).
An indecomposable generalized Cartan matrix A is said to be of finite type if it is symmetrizable and if DA is positive definite; it is said to be of affine type if it is symmetrizable and if DA is positive semidefinite of rank $n-1$; it is said to be of indefinite type if it is neither finite nor affine.
I shall now define the Kac-Moody algebra $\mathfrak{g}(A)$ associated to A and shortly recall some important results concerning $\mathfrak{g}(A)$.

Definition 2.
Given a generalized Cartan matrix $A = (a_{i,j})_{i,j\in I} \in \mathfrak{M}_{n\times n}(\mathbb{Z})$, a complex Lie algebra $\mathfrak{g}(A)$ (called the Kac-Moody algebra of A) can be associated to A in the following way:
$\mathfrak{g}(A)$ is the complex Lie algebra with generators $\{e_i, f_i, h_i|i = 1, ..., n\}$ and relations

$$[h_i, h_j] = 0,\ [h_i, e_j] = a_{i,j}e_j,\ [h_i, f_j] = -a_{i,j}f_j,\ [e_i, f_j] = \delta_{ij}h_i \ \ \forall i,j\in I,$$

$$(\mathrm{ad}e_i)^{1-a_{i,j}}(e_j) = 0,\ (\mathrm{ad}f_i)^{1-a_{i,j}}(f_j) = 0 \ \ \forall i,j\in I \text{ s. t. } i \neq j.$$

It is well known that if $\mathfrak{n}_+$, $\mathfrak{n}_-$ and $\mathfrak{h}$ are the Lie subalgebras of $\mathfrak{g}$ generated respectively by $\{e_i|i\in I\}$, $\{f_i|i\in I\}$ and $\{h_i|i\in I\}$, then $\mathfrak{n}_+$ is the Lie algebra generated by $\{e_1, ..., e_n\}$ with relations

$$(\mathrm{ad}e_i)^{1-a_{i,j}}(e_j) = 0 \quad \forall i,j = 1, ..., n \ \text{ with } i \neq j$$

and $\mathfrak{n}_-$ is the Lie algebra generated by $\{f_1, ..., f_n\}$ with relations

$$(\mathrm{ad} f_i)^{1-a_{i,j}}(f_j) = 0 \quad \forall i,j = 1,...,n \text{ with } i \neq j,$$

while $\mathfrak{h}$ is the commutative Lie algebra of dimension n.
Moreover the following theorem holds (see Kac [12]):

Theorem 3. (Triangular decomposition of $\mathfrak{g}$)
$\mathfrak{g} = \mathfrak{n}_- \oplus \mathfrak{h} \oplus \mathfrak{n}_+$.
Remark that the triangular decomposition of $\mathfrak{g}$ induces a triangular decomposition of the enveloping algebra $\mathcal{U} \doteqdot \mathcal{U}(\mathfrak{g})$ in the following way:

$$\mathcal{U} \cong \mathcal{U}^- \otimes_{\mathbb{C}} \mathcal{U}^0 \otimes_{\mathbb{C}} \mathcal{U}^+$$

where

$$\mathcal{U}^- \doteqdot \mathcal{U}(\mathfrak{n}_-), \quad \mathcal{U}^0 \doteqdot \mathcal{U}(\mathfrak{h}), \quad \mathcal{U}^+ \doteqdot \mathcal{U}(\mathfrak{n}_+).$$

□

Remark 4.
If A and B are two generalized Cartan matrices and $C \doteqdot \begin{pmatrix} A & 0 \\ 0 & B \end{pmatrix}$, then $\mathfrak{g}(C) = \mathfrak{g}(A) \oplus \mathfrak{g}(B)$ (direct sum of Lie algebras).
When A is of finite type $\mathfrak{g}(A)$ is a finite dimensional semisimple Lie algebra (simple if A is indecomposable), and, viceversa, every finite dimensional simple Lie algebra is the Kac-Moody algebra of some indecomposable Cartan matrix of finite type. On the other hand $\mathfrak{g}(A)$ is not finite dimensional if A is not finite.

I can now describe a diagram associated to A, called the Dynkin diagram of A, as follows:

Definition 5.
The Dynkin diagram associated to A is defined in the following way:
I is the set of vertices; there is an edge between $i, j \in I$ if and only if $a_{i,j} < 0$; more particularly i and j are connected by $\max(|a_{i,j}|, |a_{j,i}|)$ edges, with an arrow from i to j if $|a_{i,j}| < |a_{j,i}|$.
Notice that A is indecomposable if and only if its Dynkin diagram is connected.

Definition 6.
An automorphism of the Dynkin diagram is a permutation $\tau : I \to I$ such that $a_{\tau(i),\tau(j)} = a_{i,j}$ $\forall i, j \in I$; the set of Dynkin diagram automorphisms is naturally a group, which will be denoted by $\mathcal{T}$.
Remark that each $\tau \in \mathcal{T}$ induces an automorphism of $\mathfrak{g}$ given by

$$e_i \mapsto e_{\tau(i)}, \quad f_i \mapsto f_{\tau(i)}, \quad h_i \mapsto h_{\tau(i)} \quad \forall i \in I$$

because the relations defining $\mathfrak{g}$ depend only on the matrix A.

There is a classification of the Dynkin diagrams of finite and affine type: to each Dynkin diagram of affine type it is possible to associate one of finite type, and this correspondence is a bijection if we restrict to untwisted affine diagrams.
At the end of this part I'll list the finite and the untwisted affine Dynkin diagrams, stressing the correspondence among the two families.

In the following I suppose A to be symmetrizable.

Definition 7.
Given an indecomposable symmetrizable generalized Cartan matrix

$$A = (a_{i,j})_{i,j \in I},$$

the root lattice Q is defined as $Q \doteqdot \oplus_{i \in I} \mathbb{Z}\alpha_i \subset E \doteqdot \oplus_{i \in I} \mathbb{R}\alpha_i$.
For each element $\alpha \in E$ the height of α will be defined by $\mathrm{ht}(\alpha) \doteqdot \sum_{i \in I} |r_i|$ where $\alpha = \sum_{i \in I} r_i \alpha_i$.
The matrix DA defines on Q (resp. on E) a $\mathbb{Z}$-valued (resp. $\mathbb{R}$-valued) symmetric bilinear form $(\cdot|\cdot)$, by $(\alpha_i|\alpha_j) \doteqdot d_i a_{i,j}$.
Moreover $Q_+ \doteqdot \sum_{i \in I} \mathbb{N}\alpha_i$ defines an ordering on E by $\alpha \geq \beta \Leftrightarrow \alpha - \beta \in Q_+$.
Notice that A is of finite type if and only if $(\cdot|\cdot)$ is non-degenerate and positive.

Definition 8.
$\forall i \in I$, s_i denotes the linear transformation of E defined by

$$s_i(\alpha_j) \doteqdot \alpha_j - a_{i,j}\alpha_i.$$

Remark that $s_i^2 = \mathrm{id}$ and $s_i(Q) = Q$.
Moreover, with respect to $(\cdot|\cdot)$, s_i is a reflection of E (the reflection along α_i), because $a_{i,j} = \dfrac{(\alpha_i|\alpha_j)}{d_i} = \dfrac{2(\alpha_i|\alpha_j)}{(\alpha_i|\alpha_i)}$.
In particular s_i is an isometry of $(E, (\cdot|\cdot))$.
The group of transformations of E (of Q) generated by $\{s_1, ..., s_n\}$ is called the Weyl group of A and is denoted by W.
Notice that if $\alpha = w(\alpha_i)$ $(w \in W, i = 1, ..., n)$ and s_α denotes the reflection along α, then $s_\alpha \in W$; more precisely $s_\alpha = w s_i w^{-1}$: indeed

$$w s_i w^{-1}(\beta) = w\Big(w^{-1}(\beta) - \frac{2(\alpha_i|w^{-1}(\beta))}{(\alpha_i|\alpha_i)}\alpha_i\Big) =$$

$$= \beta - \frac{2(w(\alpha_i)|\beta)}{(w(\alpha_i)|w(\alpha_i))} w(\alpha_i) = s_\alpha(\beta) \quad \forall \beta \in E.$$

Proposition 9.
W is the Coxeter group generated by $\{s_1, ..., s_n\}$ with relations

$$s_i^2 = \mathrm{id}, \quad (s_i s_j)^{m_{ij}} = \mathrm{id} \ \text{ if } i \neq j$$

where m_{ij} is related to A in the following way ($i \neq j$):

$$a_{i,j}a_{j,i} = 0 \Rightarrow m_{ij} = 2,$$
$$a_{i,j}a_{j,i} = 1 \Rightarrow m_{ij} = 3,$$
$$a_{i,j}a_{j,i} = 2 \Rightarrow m_{ij} = 4,$$
$$a_{i,j}a_{j,i} = 3 \Rightarrow m_{ij} = 6,$$
$$a_{i,j}a_{j,i} > 3 \Rightarrow m_{ij} = \infty.$$

Proof: See Kac [12]. □

Remark that the Weyl group of A is finite if and only if A is of finite type. Moreover W is obviously a group of isometries of E.
Notice also that if $\tau \in \mathcal{T}$, τ induces an endomorphism of E setting, $\forall \tau \in \mathcal{T}$, $\tau(\alpha_i) \doteqdot \alpha_{\tau(i)}$ $\forall i \in I$. Moreover, since τ is an automorphism of the Dynkin diagram and $(\cdot|\cdot)$ depends only on A, the endomorphism induced by τ is actually an isometry.
This allows to define the extended Weyl group $\tilde{W}$ as the group of isometries of E generated by W and $\mathcal{T}$.
Remark that, as groups of isometries of E, $W \cap \mathcal{T} = \{\text{id}\}$.

Proposition 10.
$\tilde{W} \doteqdot \mathcal{T} \ltimes W$, that is

$$\tilde{W} = \{(\tau, w) = \tau w | \tau \in \mathcal{T}, w \in W\}$$

with $(\tau, w)(\tilde{\tau}, \tilde{w}) = (\tau\tilde{\tau}, \tilde{\tau}^{-1}(w)\tilde{w})$ $(\tau(s_i) \doteqdot s_{\tau(i)})$.
In particular $W \trianglelefteq \tilde{W}$, $\mathcal{T} \leq \tilde{W}$ and $\tau s_i = s_{\tau(i)}\tau$.

Proof: See Bourbaki [4]. □

Definition 11.
On W the function $l : W \to \mathbb{N}$ is defined by

$$l(w) \doteqdot \min\{r \in \mathbb{N} | \exists i_1, ..., i_r \in I \text{ s. t. } w = s_{i_1} \cdot ... \cdot s_{i_r}\};$$

$l(w)$ is called the length of w.

Proposition 12.
The function l has the following properties:
1) $l(w^{-1}) = l(w)$ $\forall w \in W$;
2)$l(s_{i_1} \cdot ... \cdot s_{i_r}) \equiv r \pmod{2}$;
3) $l(ws_i) = l(w) \pm 1$ $\forall w \in W$ $\forall i = 1, ..., n$.
Moreover the notion of length can be extended to $\tilde{W}$ by setting $l(\tau) \doteqdot 0$ $\forall \tau \in \mathcal{T}$. Notice that $l(\tau w) = l(w)$ $\forall w \in \tilde{W}$ $\forall \tau \in \mathcal{T}$. □

It is now possible to introduce the notion of root.

Definition 13.
$\forall\alpha > 0$, $\alpha = \sum_{i\in I} m_i\alpha_i$, let $\mathfrak{g}_\alpha$ and $\mathfrak{g}_{-\alpha}$ be the $\mathbb{C}$-linear subspaces of $\mathfrak{g}$ spanned respectively by

$$\{[e_{i_1},...,[e_{i_{r-1}},e_{i_r}]...]|\alpha_{i_1}+...+\alpha_{i_r}=\alpha\}$$

and

$$\{[f_{i_1},...,[f_{i_{r-1}},f_{i_r}]...]|\alpha_{i_1}+...+\alpha_{i_r}=\alpha\}.$$

Moreover define $\mathfrak{g}_0 \doteqdot \mathfrak{h}$.
Finally, if $\alpha\in Q\backslash(Q_+\cup(-Q_+))$, define $\mathfrak{g}_\alpha \doteqdot \{0\}$.
An element $\alpha\in Q$ is said to be a root if $\mathfrak{g}_\alpha \neq \{0\}$ and $\alpha\neq 0$. R denotes the set of roots. Furthermore define the set of positive roots R_+ and the set of negative roots R_- as $R_+ \doteqdot R\cap Q_+$, $R_- \doteqdot R\cap(-Q_+)$. α_1,...,α_n are called simple roots.

Proposition 14.
The root system R has the following properties:
1) $R\subset Q_+\cup(-Q_+)$, that is $R=R_+\cup R_-$;
2) $R_- = -R_+$;
3) $m\alpha_i\in R \Leftrightarrow m=\pm 1$;
4) $\dim_{\mathbb{C}}\mathfrak{g}_{\alpha_i} = 1\ \ \forall i\in I$.
Moreover $\mathfrak{g}$ is Q-graded and the following facts hold:
1) $\mathfrak{n}_+ = \bigoplus_{\alpha>0}\mathfrak{g}_\alpha = \bigoplus_{\alpha\in R_+}\mathfrak{g}_\alpha$ and $\mathfrak{n}_- = \bigoplus_{\alpha>0}\mathfrak{g}_{-\alpha} = \bigoplus_{\alpha\in R_+}\mathfrak{g}_{-\alpha}$;
2) $\mathfrak{g} = \bigoplus_{\alpha\in Q}\mathfrak{g}_\alpha = \left(\bigoplus_{\alpha\in R}\mathfrak{g}_\alpha\right)\oplus\mathfrak{h}$;
3) $\forall\alpha,\beta\in Q$ $[\mathfrak{g}_\alpha,\mathfrak{g}_\beta]\subseteq\mathfrak{g}_{\alpha+\beta}$;
4) $\forall\alpha\in Q$ $\mathfrak{g}_\alpha$ is finite dimensional over $\mathbb{C}$;
5) $\forall\alpha\in Q$, $\forall x\in\mathfrak{g}_\alpha$, $\forall h\in\mathfrak{h}$, $[h,x]=\alpha(h)x$ where $\forall i=1,...,n$ α_i is the linear function on $\mathfrak{h}$ defined by $\alpha_i(h_j)\doteqdot a_{j,i}$ $\forall j=1,...,n$.

Proof: See Kac [12]. □

R also has good properties with respect to the action of the Weyl group W; indeed since ade_i and adf_i are locally nilpotent on $\mathfrak{g}$ one can define exp($\pm$ade_i) and exp($\pm$adf_i), which are automorphisms of $\mathfrak{g}$. Then

$$\tau_i \doteqdot \exp(\mathrm{ad}e_i)\exp(-\mathrm{ad}f_i)\exp(\mathrm{ad}e_i)$$

is again an automorphism of $\mathfrak{g}$ and, calculating $\tau_i(e_j)$ and $\tau_i(f_j)$, one immediately sees that $\tau_i(\mathfrak{g}_\alpha)\subseteq\mathfrak{g}_{s_i(\alpha)}$ $\forall\alpha\in Q$.
This implies the following proposition:

Proposition 15.
R is W-stable.
More particularly $\forall\alpha\in R$, $\dim_{\mathbb{C}}\mathfrak{g}_{w(\alpha)} = \dim_{\mathbb{C}}\mathfrak{g}_\alpha$ $\forall w\in W$ (the dimension of $\mathfrak{g}_\alpha$ is called multiplicity of the root α). □

We are now able to study more closely the relations between R and W.

Proposition 16.
1) Let $\alpha \in R_+$, $i \in I$; then $s_i(\alpha) < 0$ if and only if $\alpha = \alpha_i$. In particular $s_i(R_+ \setminus \{\alpha_i\}) = R_+ \setminus \{\alpha_i\}$.
2) $\forall w \in W$, $\forall i = 1, ..., n$,

$$l(ws_i) = l(w) + 1 \Leftrightarrow w(\alpha_i) > 0.$$

Proof: See Bourbaki [4] and Kac [12]. □

Proposition 17.
Let $w = s_{i_1} \cdot ... \cdot s_{i_r}$ be a reduced expression; then

$$\{s_{i_1} \cdot ... \cdot s_{i_{t-1}}(\alpha_{i_t}) | t = 1, ..., r\} = \{\alpha \in R_+ | w^{-1}(\alpha) < 0\}.$$

In particular $\#(R_+ \cap w(R_-)) = l(w)$.
Proof: On one hand

$$w^{-1}s_{i_1} \cdot ... \cdot s_{i_{t-1}}(\alpha_{i_t}) = s_{i_r} \cdot ... \cdot s_{i_t}(\alpha_{i_t}) = -s_{i_r} \cdot ... \cdot s_{i_{t+1}}(\alpha_{i_t}) < 0.$$

On the other hand if $w^{-1}(\alpha) < 0$ $\exists t \in \{1, ..., r\}$ such that $s_{i_t} \cdot ... \cdot s_{i_1}(\alpha) < 0$ and $s_{i_{t-1}} \cdot ... \cdot s_{i_1}(\alpha) > 0$, that is $s_{i_{t-1}} \cdot ... \cdot s_{i_1}(\alpha) = \alpha_{i_t}$; then $\alpha = s_{i_1} \cdot ... \cdot s_{i_{t-1}}(\alpha_{i_t})$.
□

The action of W on R allows also to define the notions of real and imaginary roots:

Definition 18.
Real and imaginary roots are defined by:
$R^{\text{re}} \doteqdot W(\{\alpha_i | i \in I\})$, that is, a root is real if and only if it is conjugate to some simple root; $R^{\text{im}} \doteqdot R \setminus R^{\text{re}}$, that is a root is imaginary if it is not real.
I also define $R_+^{\text{re}} \doteqdot R_+ \cap R^{\text{re}}$ and $R_+^{\text{im}} \doteqdot R_+ \cap R^{\text{im}}$.

Proposition 19.
$R^{\text{re}} = \{\alpha \in R | (\alpha|\alpha) > 0\}$, $R^{\text{im}} = \{\alpha \in R | (\alpha|\alpha) \leq 0\}$.
In particular if A is of finite type every root is real (and R is finite, since W is), while if A is affine $R^{\text{im}} = \{\alpha \in Q | (\alpha|\alpha) = 0\} = Q \cap \ker(\cdot|\cdot) \setminus \{0\}$, so that every imaginary root is fixed by W (and also by $\mathcal{T}$, because $\ker(\cdot|\cdot)$ is 1-dimensional) and R is not finite.
R^{re} is finite if and only if A is finite.
Proof: See Bourbaki [4], and Kac [12]. □

Remark 20.
$\dim(\mathfrak{g}_\alpha) = 1 \quad \forall \alpha \in R^{\text{re}}$. □

The multiplicity of the imaginary roots is generally very complicated, and not known in the generality of cases. However in the affine case the dimension of the root spaces is known and I'll describe it briefly for the untwisted affine algebras in the next paragraphs.

I end this paragraph with some definitions and results concerning the dimension of $\mathcal{U}_\eta$ for $\eta\in Q$: indeed notice that, as a consequence of the theorem of Poincaré-Birkhoff-Witt for Lie algebras, the enveloping algebra $\mathcal{U}$ of $\mathfrak{g}$ is a Q-graded algebra:

$$\mathcal{U}=\oplus_{\alpha\in Q}\mathcal{U}_\alpha$$

where $\forall\alpha\in Q$ $\mathcal{U}_\alpha$ is the $\mathbb{C}$-linear span of the vectors

$$f_{i_1}\cdot\ldots\cdot f_{i_r}h_{j_1}\cdot\ldots\cdot h_{j_s}e_{k_1}\cdot\ldots\cdot e_{k_t}$$

such that $\alpha_{k_1}+\ldots+\alpha_{k_t}-(\alpha_{i_1}+\ldots+\alpha_{i_r})=\alpha$; in particular

$$\mathcal{U}_\alpha=\oplus_{\beta\in Q}\mathcal{U}^-_{-\beta}\mathcal{U}^0\mathcal{U}^+_{\alpha+\beta}$$

where, if $\eta\in Q$, $\mathcal{U}^\pm_\eta\doteq\mathcal{U}_\eta\cap\mathcal{U}^\pm$.

Definition 21.
I define $\tilde{R}$ to be the set of roots with multiplicity, that is

$$\tilde{R}\doteq\{(\alpha,r)\in R\times\mathbb{Z}_+|1\le r\le\dim_{\mathbb{C}}\mathfrak{g}_\alpha\};$$

$p:\tilde{R}\to R$ denotes the natural projection on R, that is $p(\alpha,r)\doteq\alpha$ $\forall(\alpha,r)\in\tilde{R}$. Set $\tilde{R}_+\doteq p^{-1}(R_+)$.

Definition 22.
$\forall\eta\in Q$ define $\mathrm{Par}(\eta)$ to be

$$\mathrm{Par}(\eta)\doteq\{(\gamma_1,\ldots,\gamma_r)|r\in\mathbb{N},\ \gamma_i\in\tilde{R}_+\ \forall i,\sum_{i=1}^r p(\gamma_i)=\eta\}/\sim$$

where $\sim$ is the equivalence relation in $\cup_{r\in\mathbb{N}}(\tilde{R}_+)^r$ given by

$$(\gamma_1,\ldots,\gamma_r)\sim(\gamma'_1,\ldots,\gamma'_{r'})\text{ if and only if}$$

$$r=r'\text{ and }\exists\sigma\in\mathfrak{S}_r\text{ such that }\gamma'_i=\gamma_{\sigma(i)}\quad\forall i=1,\ldots,r.$$

Equivalently $\mathrm{Par}(\eta)=\{(\gamma_1\preceq\ldots\preceq\gamma_r)|r\in\mathbb{N},\ \gamma_i\in\tilde{R}_+\ \forall i,\sum_{i=1}^r p(\gamma_i)=\eta\}$ where $\preceq$ is any linear ordering of $\tilde{R}_+$.
Also define $\mathrm{par}(\eta)$ to be the cardinality of $\mathrm{Par}(\eta)$.

Proposition 23.
$\forall\eta\in Q$, $\mathrm{par}(\eta)<\infty$ and $\dim_{\mathbb{C}}\mathcal{U}^+_\eta=\dim_{\mathbb{C}}\mathcal{U}^-_{-\eta}=\mathrm{par}(\eta)$.

Proof: The first assertion follows from the fact that $\dim\mathfrak{g}_\alpha<\infty$ $\forall\alpha\in Q$; the second one immediately follows from the theorem of Poincaré-Birkhoff-Witt.
□

§A2. Finite case: coroots and coweights.
Suppose now that A is finite and indecomposable: in this case the non degeneracy of A allows us to define some more structure associated to A. Before introducing coroots and coweights I recall the following fact:

Proposition 1.
The root system is finite, and so is the Weyl group; in particular there exists a unique maximal root, denoted by θ; θ is called the highest root and is such that
1) $\theta \geq \alpha \ \ \forall \alpha \in R$;
2) $(\theta|\theta) \geq (\alpha|\alpha) \ \ \forall \alpha \in R$.

Proof: See Bourbaki [4]. □

Definition 2.
The coroot system $R\check{}$ of A is defined by $R\check{} \doteqdot \{\alpha\check{} \in E^*|\alpha \in R\}$ where $\forall \alpha \in R$ $\alpha\check{} \in E^*$ is such that $\langle \alpha\check{}, \alpha\rangle = 2$ and $\beta - \langle \alpha\check{}, \beta\rangle\alpha \in R \ \ \forall \beta \in R$. Each $\alpha\check{} \in E^*$ is said to be a coroot.
The first aim of this paragraph is to show that $R\check{}$ is well defined. To this aim one needs the following lemma:

Lemma 3.
Let $\alpha \in R$ and $f, g \in E^*$ be such that $f(\alpha) = g(\alpha) = 2$ and

$$\beta - f(\beta)\alpha, \beta - g(\beta)\alpha \in R \ \ \forall \beta \in R;$$

then $f = g$.

Proof: See Bourbaki [4]. □

Corollary 4.
Let $\alpha \in R$; if $\exists \alpha\check{} \in E^*$ such that $\langle \alpha\check{}, \alpha\rangle = 2$ and $\beta - \langle \alpha\check{}, \beta\rangle\alpha \in R \ \forall \beta \in R$, then $\alpha\check{}$ is the unique element of E^* satisfying these properties. □

For following use recall that W acts on E^* $(w(f) \doteqdot f \circ w^{-1} \ \forall f \in E^*)$.

Proposition 5.
Suppose that there exist the elements $\alpha_i\check{}$ for $i \in I$. Then, $\forall \alpha \in R$, $\alpha\check{}$ exists, and it is described as follows: if $\alpha = w(\alpha_i)$ with $w \in W$, $i \in I$, then $\alpha\check{} = w(\alpha_i\check{})$. In particular $w(\alpha_i\check{})$ does not depend on w, i but just on α.

Proof: It is an immediate calculation, see Bourbaki [4]. □

I prove now that $\alpha_i\check{}$ exists $\forall i \in I$, giving a more precise description of $\{\alpha_1\check{}, ..., \alpha_n\check{}\}$.

Definition 6.
$\{\omega_1\check{}, ..., \omega_n\check{}\}$ denotes the dual basis of E^* with respect to $\{\alpha_1, ..., \alpha_n\}$, that is

$$\langle \omega_i\check{}, \alpha_j\rangle = \delta_{ij} \ \ \forall i, j = 1, ..., n;$$

$P^\check{}$ denotes the lattice in E^* generated by $\{\omega_1^\check{}, ..., \omega_n^\check{}\}$: $P^\check{} = \bigoplus_{i\in I} \mathbb{Z}\omega_i^\check{}$; the elements of $P^\check{}$ are called coweights.
Remark that $W(P^\check{}) = P^\check{}$: more precisely $s_i(\omega_j^\check{}) = \omega_j^\check{} - \delta_{ij} \sum_{k\in I} a_{i,k}\omega_k^\check{}$.
Indeed

$$\langle s_i(\omega_j^\check{}), \alpha_k\rangle = \langle \omega_j^\check{}, s_i(\alpha_k)\rangle = \langle \omega_j^\check{}, \alpha_k - a_{i,k}\alpha_i\rangle = \delta_{jk} - a_{i,k}\delta_{ij}.$$

Since $s_i^2 = \mathrm{id}$ this implies that $W(P^\check{}) = P^\check{}$.

Proposition 7.
$\exists \alpha_1^\check{}, ..., \alpha_n^\check{} \in P^\check{}$ (from which $R^\check{} \subseteq P^\check{}$); more exactly $\forall i\in I$ $\alpha_i^\check{} = \sum_{j\in I} a_{i,j}\omega_j^\check{}$ and $\beta - \langle \alpha_i^\check{}, \beta\rangle\alpha_i = s_i(\beta)$ $\forall \beta\in E$.
In particular $s_i(\omega_j^\check{}) = \omega_j^\check{} - \delta_{ij}\alpha_i^\check{}$ and $\langle \alpha_i^\check{}, \alpha_j\rangle = a_{ij}$ $\forall i,j\in I$.

Proof: It is a matter of calculation, and is immediate. □

Notice that if $s_\alpha \in W$ is the reflection along $\alpha = w(\alpha_i)$ then $\forall \beta\in E$

$$s_\alpha(\beta) = ws_iw^{-1}(\beta) = w(w^{-1}(\beta) - \langle \alpha_i^\check{}, w^{-1}(\beta)\rangle\alpha_i) = \beta - \langle \alpha^\check{}, \beta\rangle\alpha;$$

hence $\forall \alpha, \beta\in R$ $\langle \alpha^\check{}, \beta\rangle = \dfrac{2(\alpha|\beta)}{(\alpha|\alpha)}$, which means that under the identification between E and E^* induced by $(\cdot|\cdot)$ we have that $\alpha^\check{} = \dfrac{2\alpha}{(\alpha|\alpha)}$.
In particular $(\alpha|\alpha) \leq (\beta|\beta) \Rightarrow (\alpha^\check{}|\alpha^\check{}) \geq (\beta^\check{}|\beta^\check{})$.
Moreover, since A is non degenerate, proposition 1.A2.7 implies that

$$\{\alpha_1^\check{}, ..., \alpha_n^\check{}\} \text{ is a basis of } E^*.$$

Remark that $R^\check{} = R({}^tA)$ because

$$s_i(\alpha_j^\check{}) = \sum_{k\in I} a_{j,k}(\omega_k^\check{} - \delta_{ik}\alpha_i^\check{}) = \alpha_j^\check{} - a_{j,i}\alpha_i^\check{}.$$

Then $R^\check{} \subset Q^\check{} \doteq Q({}^tA) = \bigoplus_{i\in I} \mathbb{Z}\alpha_i^\check{} \subseteq P^\check{}$.
Also, $P^\check{}$ and $Q^\check{}$ act on E^* as groups of translations and preserve $P^\check{}$ ($Q^\check{}$ preserves $Q^\check{}$, too). As groups of transformations of E^*, we have that

$$Q^\check{} \cap W = P^\check{} \cap W = \{\mathrm{id}\}$$

and $Q^\check{}, P^\check{}$ are normal in the group generated by $P^\check{}$ and W, because

$$wxw^{-1} = w(x) \ \ \forall w\in W,\ x\in P^\check{}.$$

Then the group generated by $P^\check{}$ and W is $W \ltimes P^\check{}$ and the group generated by $Q^\check{}$ and W is $W \ltimes Q^\check{}$, and both are groups of transformations of E^* preserving $P^\check{}$ ($W \ltimes Q^\check{}$ preserves also $Q^\check{}$).

Lemma 8.
Let $\alpha \in R$; then $W.\alpha = \{\beta \in R | (\beta|\beta) = (\alpha|\alpha)\}$.
If $\alpha \in R_+$ is of minimal length, that is $(\alpha|\alpha) \le (\beta|\beta)$ $\forall \beta \in R$, then the conjugates of α, that is $\{\beta \in R | (\alpha|\alpha) = (\beta|\beta)\}$, generate Q.

Proof: For the first assertion see Bourbaki [4].
For the second assertion: if all the roots have the same length the claim is obvious. Otherwise $\#\{(\beta|\beta) | \beta \in R\} = 2$ and it suffices to prove that $\exists i \in I$ such that $(\alpha_i|\alpha_i) > (\alpha|\alpha)$ and α_i belongs to the $\mathbb{Z}$-lattice generated by $\{\beta \in R | (\beta|\beta) = (\alpha|\alpha)\}$.
Let $i \in I$ be such that $(\alpha_i|\alpha_i)$ is maximal and $\exists j \in I$ such that $(\alpha_j|\alpha_j) = (\alpha|\alpha)$ with $a_{i,j} \neq 0$; then $d_j < d_i$, that is $|a_{j,i}| > |a_{i,j}| = 1$, so that

$$s_i(\alpha_j) - \alpha_j = -a_{i,j}\alpha_i = \alpha_i,$$

which proves the thesis. □

I shall now introduce the special element $\rho\check{}$, which will play an important role in the following:

Definition 9.
$\rho\check{} \doteqdot \frac{1}{2}\sum_{\alpha \in R_+} \alpha\check{}$.
It is well known that $\rho\check{} = \sum_{i \in I} \omega_i\check{}$.
Define also $P_+\check{} \doteqdot \sum_{i \in I} \mathbb{N}\omega_i\check{}$ and $P_{++}\check{} \doteqdot \{\sum_{i \in I} m_i\omega_i\check{} | m_i > 0 \ \forall i \in I\} = \rho\check{} + P_+\check{}$.
The elements of $P_+\check{}$ and $P_{++}\check{}$ are said to be respectively dominant and strongly dominant coweights.

§A3. Untwisted affine case.

Consider now the case when A is a generalized Cartan matrix of untwisted affine type (and it is indecomposable). In this case there is, associated with A, an indecomposable Cartan matrix A_0 of finite type, where $A_0 = (a_{i,j})_{i,j \in I_0}$ ($I_0 \doteqdot I \setminus \{r\}$ where r is a suitable element of I: in the following I'll say that $I = \{0, ..., n\}$ and $I_0 = \{1, ..., n\}$). In this case, if we set $\mathfrak{g} \doteqdot \mathfrak{g}(A_0)$, $\hat{\mathfrak{g}} \doteqdot \mathfrak{g}(A)$, it is well known that $\hat{\mathfrak{g}} = \mathfrak{g} \otimes \mathbb{C}[t, t^{-1}] \oplus \mathbb{C}c$ is a central extension of $\mathfrak{g} \otimes \mathbb{C}[t, t^{-1}]$. The structures illustrated till now relative to A and A_0 have important relations: in particular it is possible to describe W ($\tilde{W}$: see proposition 1.A1.10) in terms of the coroots (coweights) and W_0 (Weyl group of the associated finite algebra) and analogously R in terms of R_0 (root system of the associated finite algebra).

Remark 1.
There is an injection $Q_0 (\doteqdot Q(A_0)) \hookrightarrow Q$; $Q = Q_0 \oplus \mathbb{Z}\delta$ where $\delta = \theta + \alpha_0 \in Q_+$ is such that $\mathbb{Z}\delta = Q \cap \ker(\cdot|\cdot)$ (θ is the highest root of A_0).
Notice also that $w(\delta) = \delta$ $\forall w \in \tilde{W}$ and that $W_0 \doteqdot W(A_0) \subset W$ preserves Q_0.

Definition 2.
Define an action of $P_0\check{} = P\check{}(A_0)$ on Q by $x(\alpha) \doteqdot \alpha - \langle x, \alpha \rangle \delta$ if $\alpha \in Q_0$ and $x(\delta) \doteqdot \delta$.

Remark that $\forall x \in P_0^\vee$ $\alpha \mapsto x(\alpha)$ is an isometry of E.
As groups of isometries of E we have $P_0^\vee \cap W_0 = \{\mathrm{id}\}$; furthermore the fact that $\forall w \in W_0$ $\forall x \in P_0^\vee$

$$wxw^{-1}(\alpha) = w(w^{-1}(\alpha) - \langle x, w^{-1}(\alpha)\rangle\delta) = \alpha - \langle w(x), \alpha\rangle\delta = w(x)(\alpha) \quad \forall \alpha \in E_0,$$

$$wxw^{-1}(\delta) = \delta = w(x)(\delta)$$

means that $wxw^{-1} = w(x)$ on E, that is $W_0 \ltimes P_0^\vee$ and $W_0 \ltimes Q_0^\vee$ act on E and both fix Q. Finally, the group homomorphism from $W_0 \ltimes P_0^\vee$ into the group of (affine) isometries of E is injective.
I want now to prove that $W_0 \ltimes Q_0^\vee \cong W$ and $W_0 \ltimes P_0^\vee \cong \tilde{W}$.

Lemma 3.
$s_0 = s_\theta \circ (-\theta^\vee)$ as transformations of E, hence in $W_0 \ltimes Q_0^\vee$.
This means that $s_0 \in W_0 \ltimes Q_0^\vee$; more generally $W \subseteq W_0 \ltimes Q_0^\vee$.

Proof: Since both s_0 and $s_\theta \circ (-\theta^\vee)$ leave δ fixed, it is enough to show that they are the same on E_0: this is just an easy computation. □

Proposition 4.
$W = W_0 \ltimes Q_0^\vee$ and $\tilde{W} \cong W_0 \ltimes P_0^\vee$.

Proof: Obviously $W_0 \subset W$; moreover $\theta^\vee = s_0 s_\theta \in W$, so that the linear combinations with integer coefficients of the W_0-conjugates of $\theta^\vee$ are in W (because $w(\theta^\vee) = w\theta^\vee w^{-1}$). But the set of W_0-conjugates of $\theta^\vee$ is the set of the roots in $R^\vee$ with the same length as $\theta^\vee$, that is of minimal length. Then lemma 1.A2.8 implies that $Q_0^\vee \subset W$, and the thesis follows.
The second assertion follows from the fact that $P_0^\vee / Q_0^\vee \cong \mathcal{T}$, see Bourbaki [4].
□

I shall now recall some properties of $P_0^\vee \triangleleft \tilde{W}$ (see Beck [1] and Bourbaki [4]):
$\forall x \in P_{0,+}^\vee$ $l(s_i x) = l(x) + 1$ $\forall i \in I_0$, that is any reduced expression of x starts with s_0.
Moreover $\forall x, y \in P_{0,+}^\vee$, $l(xy) = l(x) + l(y)$.
Finally $\forall i, j \in I_0$ $l(\omega_i^\vee s_j) = l(\omega_i^\vee) + (-1)^{\delta_{ij}}$, that is any reduced expression of $\omega_i^\vee$ starts with s_0 and ends with s_i.

At last I describe the root system R.

Proposition 5.
Real and imaginary roots are given by

$$R^{\mathrm{re}} = R_0 + \mathbb{Z}\delta = \{\alpha + m\delta \mid \alpha \in R_0, m \in \mathbb{Z}\}, \quad R^{\mathrm{im}} = \mathbb{Z}\delta \setminus \{0\}.$$

Moreover $\dim_{\mathbb{C}} \mathfrak{g}_{m\delta} = \# I_0 = n$ $\forall m \in \mathbb{Z} \setminus \{0\}$.

Proof: It follows from the fact that

$$\hat{\mathfrak{g}} \cong \mathfrak{g} \otimes \mathbb{C}[t, t^{-1}] \oplus \mathbb{C}c = (\mathfrak{h} \oplus \mathbb{C}c) \oplus \Big(\bigoplus_{\substack{\alpha \in R_0 \\ m \in \mathbb{Z}}} \mathfrak{g}_\alpha \otimes \mathbb{C}t^m \Big) \oplus \Big(\bigoplus_{m > 0} \mathfrak{h} \otimes \mathbb{C}t^m \Big).$$

In particular $\mathfrak{g}_{m\delta} = \mathfrak{h} \otimes \mathbb{C}t^m$ (see Kac [12]). □

Remark 6.
One can identify $\tilde{R}_+$ with $R_+^{\mathrm{re}} \cup (R_+^{\mathrm{im}} \times I_0)$.

§A4. Classification of finite and untwisted affine Dynkin diagrams.

Before going on in the exposition I recall the classification of finite and untwisted affine Dynkin diagrams. In the list below the vertices of the untwisted affine diagrams are indexed by $I = \{0, 1, ..., n\}$, while the labels on the vertices of the finite type diagrams (whose indicization is induced by $I_0 = I \setminus \{0\}$) are the coefficients $r_i \doteq \langle \omega_i^\vee, \theta \rangle$ of α_i in θ (that is $\theta = \sum_{i \in I_0} r_i \alpha_i$). Recall that $\delta = \theta + \alpha_0 = \sum_{i \in I} r_i \alpha_i$ with $r_0 = 1$.

Untwisted affine type Finite type

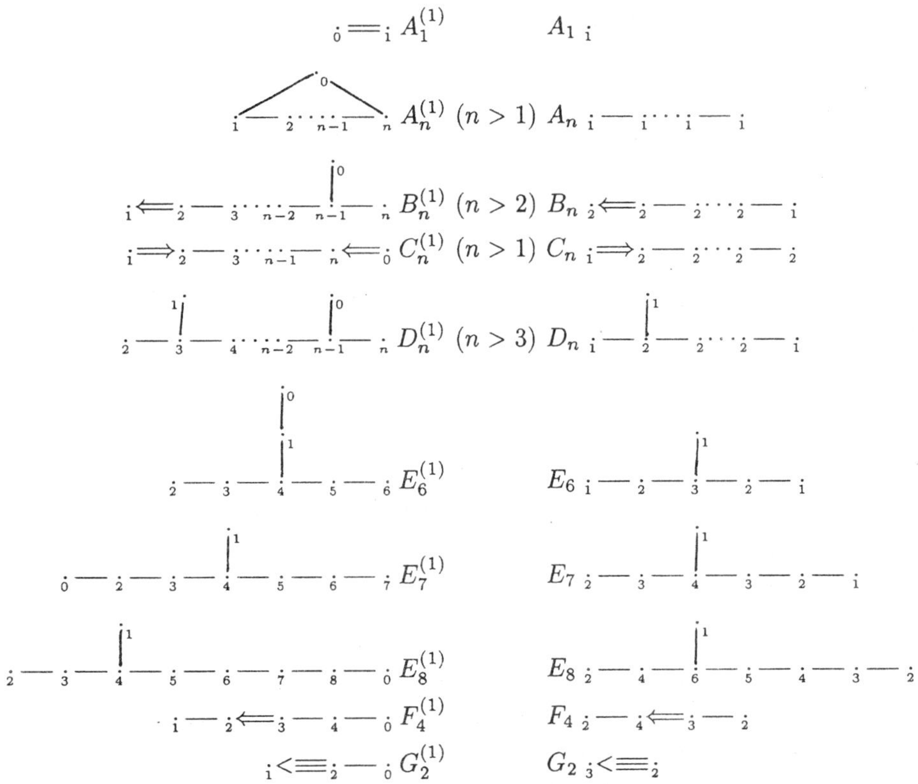
$A_1^{(1)}$
A_1
$A_n^{(1)}$ $(n > 1)$
A_n
$B_n^{(1)}$ $(n > 2)$
B_n
$C_n^{(1)}$ $(n > 1)$
C_n
$D_n^{(1)}$ $(n > 3)$
D_n
$E_6^{(1)}$
E_6
$E_7^{(1)}$
E_7
$E_8^{(1)}$
E_8
$F_4^{(1)}$
F_4
$G_2^{(1)}$
G_2

Part B. The quantum algebra.

I shall now introduce the quantum algebra associated to a symmetrizable generalized Cartan matrix. This was first introduced by Drinfeld and Jimbo for finite dimensional semisimple Lie algebras and it is a q-deformation of the enveloping algebra $\mathcal{U}$ of the Kac-Moody algebra.

§B1. Triangular decomposition and Q-gradation.

Definition 1.
Given a symmetrizable generalized Cartan matrix $A = (a_{i,j})_{i,j\in I} \in \mathfrak{M}_{n\times n}(\mathbb{Z})$, an associative algebra $\mathcal{U}_q(A)$, called the quantum algebra of A, can be associated to A in the following way: $\mathcal{U}_q(A)$ is the $\mathbb{C}(q)$-associative algebra (with 1) generated by $\{E_i, F_i, K_i, \tilde{K}_i | i\in I\}$ with relations

$$K_i\tilde{K}_i = 1 = \tilde{K}_iK_i \ (\Rightarrow \tilde{K}_i = K_i^{-1}) \ \forall i\in I,$$

$$\begin{aligned} K_iK_j &= K_jK_i \\ K_iE_j &= q_i^{a_{i,j}}E_jK_i \\ K_iF_j &= q_i^{-a_{i,j}}F_jK_i \qquad \forall i,j\in I, \\ [E_i,F_j] &= E_iF_j - F_jE_i = \delta_{ij}\frac{K_i - K_i^{-1}}{q_i - q_i^{-1}} \end{aligned}$$

$$\sum_{r=0}^{1-a_{i,j}} (-1)^r \binom{1-a_{i,j}}{r}_{q_i} E_i^rE_jE_i^{1-a_{i,j}-r} = 0$$

$$\sum_{r=0}^{1-a_{i,j}} (-1)^r \binom{1-a_{i,j}}{r}_{q_i} F_i^rF_jF_i^{1-a_{i,j}-r} = 0 \qquad \forall i,j\in I \text{ s. t. } i\neq j,$$

where $q_i \doteqdot q^{d_i}$ $\forall i\in I$, $[m]_v \doteqdot \frac{v^m - v^{-m}}{v - v^{-1}}$ $\forall m\in\mathbb{Z}$, $[m]_v! \doteqdot \prod_{r=1}^m [r]_v$ $\forall m\in\mathbb{N}$ and

$$\binom{m}{k}_v \doteqdot \frac{[m]_v!}{[k]_v![m-k]_v!} \quad \forall k\leq m\in\mathbb{N}.$$

If $\mathfrak{g} = \mathfrak{g}(A)$ I shall often write $\mathcal{U}_q(\mathfrak{g})$ (or simply $\mathcal{U}_q$) for $\mathcal{U}_q(A)$.

Remark 2.
If A and B are two generalized Cartan matrices and $C \doteqdot \begin{pmatrix} A & 0 \\ 0 & B\end{pmatrix}$, then $\mathcal{U}_q(C) = \mathcal{U}_q(A)\otimes_{\mathbb{C}(q)}\mathcal{U}_q(B)$ (tensor product of algebras). It is then enough to study the algebras associated to indecomposable matrices.
For $\mathcal{U}_q$ results analogous to those described for the enveloping algebra $\mathcal{U}$ hold; I recall the following definitions:

Definition 3.
1) $\mathcal{U}_q^+$, $\mathcal{U}_q^-$, $\mathcal{U}_q^0$, $\mathcal{U}_q^{\geq 0}$ and $\mathcal{U}_q^{\leq 0}$ are the $\mathbb{C}(q)$-subalgebras of $\mathcal{U}_q$ generated respectively by $\{E_i | i\in I\}$, $\{F_i | i\in I\}$, $\{K_i^{\pm 1} | i\in I\}$, $\{E_i, K_i^{\pm 1} | i\in I\}$ and $\{F_i, K_i^{\pm 1} | i\in I\}$;

2) $\forall\alpha\in Q$ $\mathcal{U}_{q,\alpha}$ denotes the $\mathbb{C}(q)$-linear span of the vectors of the form

$F_{i_1}\cdot\ldots\cdot F_{i_r}K_\lambda E_{j_1}\cdot\ldots\cdot E_{j_s}$ with $\lambda\in Q$ and $\alpha_{j_1}+\ldots+\alpha_{j_s}-(\alpha_{i_1}+\ldots+\alpha_{i_r})=\alpha$;

3) $\mathcal{U}_{q,\alpha}^+ \doteq \mathcal{U}_{q,\alpha}\cap\mathcal{U}_q^+$, $\mathcal{U}_{q,\alpha}^- \doteq \mathcal{U}_{q,\alpha}\cap\mathcal{U}_q^-$;
4) π and $\tilde{\pi}$ are the $\mathbb{C}(q)$-linear homomorphisms from $\mathcal{U}_q$ to $\mathcal{U}_q^0$ and $\mathcal{U}_q^{\leq 0}$ respectively defined by:

$$\pi(xE_i)=0 \text{ and } \pi(F_ix)=0\ \forall i\in I,\forall x\in\mathcal{U}_q \text{ and } \pi\big|_{\mathcal{U}_q^0}=\mathrm{id}_{\mathcal{U}_q^0};$$

$$\tilde{\pi}(xE_i)=0\quad \forall i\in I,\forall x\in\mathcal{U}_q \text{ and } \tilde{\pi}\big|_{\mathcal{U}_q^{\leq 0}}=\mathrm{id}_{\mathcal{U}_q^{\leq 0}}.$$

Theorem 4.
1) The $\mathbb{C}(q)$-linear homomorphism $\chi:\mathcal{U}_q^-\otimes\mathcal{U}_q^0\otimes\mathcal{U}_q^+\to\mathcal{U}_q$ given by

$$\chi(x\otimes y\otimes z)\doteq xyz$$

is an isomorphism of $\mathbb{C}(q)$-vector spaces (triangular decomposition);
2) $\mathcal{U}_q=\oplus_{\alpha\in Q}\mathcal{U}_{q,\alpha}$, $\mathcal{U}_q^+=\oplus_{\alpha\in Q_+}\mathcal{U}_{q,\alpha}^+$, $\mathcal{U}_q^-=\oplus_{\alpha\in Q_+}\mathcal{U}_{q,-\alpha}^-$;
3) $\forall\alpha\in Q$ $\mathcal{U}_{q,\alpha}=\oplus_{\beta\in Q_+}(\mathcal{U}_{q,-\beta}^-\mathcal{U}_q^0\mathcal{U}_{q,\alpha+\beta}^+)$;
4) in the decomposition

$$\mathcal{U}_q=\bigoplus_{\substack{\alpha,\beta\in Q\\ \beta,\alpha+\beta\in Q_+}}\mathcal{U}_{q,-\beta}^-\mathcal{U}_q^0\mathcal{U}_{q,\alpha+\beta}^+$$

π and $\tilde{\pi}$ are the projections on $\mathcal{U}_q^0$ and $\mathcal{U}_q^{\leq 0}=\oplus_{\alpha\in Q_+}\mathcal{U}_{q,-\alpha}^{\leq 0}$ respectively;
5) $\forall\alpha\in Q$ $\dim_{\mathbb{C}(q)}\mathcal{U}_{q,\alpha}^+=\dim_{\mathbb{C}(q)}\mathcal{U}_{q,-\alpha}^-=\mathrm{par}(\alpha)$.
Proof: For the proof see Lusztig [15]. □

§B2. Braid group action.

I now introduce the braid group $\mathcal{B}$ and its action on $\mathcal{U}_q$.

Definition 1.
The braid group $\mathcal{B}$ associated to W is the group generated by $\{T_i|i\in I\}$ with relations $T_{ij}=T_{ji}$ for $i,j\in I$ such that $\mathrm{ord}(s_is_j)<\infty$, where, if $\mathrm{ord}(s_is_j)<\infty$, T_{ij} is defined in the following way:

$$T_{ij}\doteq T_{i_1}\cdot\ldots\cdot T_{i_{\mathrm{ord}(s_is_j)}}\quad\text{where}\quad T_{i_r}=\begin{cases}T_i & \text{if } r \text{ odd}\\ T_j & \text{if } r \text{ even.}\end{cases}$$

Remark 2.
Remark that $1\to\mathcal{N}\to\mathcal{B}\to W\to 1$ ($\mathcal{B}\ni T_i\mapsto s_i\in W$) where $\mathcal{N}$ is the normal subgroup of $\mathcal{B}$ generated by $\{T_i^2|i\in I\}$.

Definition 3.
$T: W \to \mathcal{B}$ $(w \mapsto T_w)$ will denote the (unique) section of W in $\mathcal{B}$ such that $s_i \mapsto T_i$ and $T_{w\tilde{w}} = T_w T_{\tilde{w}}$ whenever $l(w\tilde{w}) = l(w) + l(\tilde{w})$.
Such a section exists (see Matsumoto [19]).
Moreover $\forall w \in W$ the image of T_w in W is w, but T is not a group homomorphism.
Notice also that if A is affine untwisted and $x, y \in P^{\vee}_{0,+}$, then

$$T_x T_y = T_{x+y} = T_y T_x,$$

because in $\tilde{W}$ $x + y = xy$ and $l(xy) = l(x) + l(y)$.

Definition 4.
As we did for the extended Weyl group $\tilde{W}$, it is possible to define the extended braid group $\tilde{\mathcal{B}}$ in the following way:

$$\tilde{\mathcal{B}} = \{(\tau, T) = \tau T | \tau \in \mathcal{T}, T \in \mathcal{B}\}$$

with $(\tau, T)(\tilde{\tau}, \tilde{T}) = (\tau\tilde{\tau}, \tilde{\tau}^{-1}(T)\tilde{T})$ (setting $\tau(T_i) \doteqdot T_{\tau(i)}$).

Remark 5.
$\mathcal{B} \trianglelefteq \tilde{\mathcal{B}}$ and $\mathcal{T} \leq \tilde{\mathcal{B}}$.
Moreover the section $T: W \to \mathcal{B}$ extends to a section $T: \tilde{W} \to \tilde{\mathcal{B}}$ such that $T_{w\tilde{w}} = T_w T_{\tilde{w}}$ whenever $l(w\tilde{w}) = l(w) + l(\tilde{w})$.
Notice that $T|_{\mathcal{T}}$ is the identity. □

Definition 6.
$\forall i \in I$ define the homomorphism of $\mathbb{C}(q)$-algebras $T_i : \mathcal{U}_q \to \mathcal{U}_q$ on the generators of $\mathcal{U}_q$ in the following way:

$$T_i(E_i) \doteqdot -F_i K_i, \quad T_i(E_j) \doteqdot \sum_{r=0}^{-a_{i,j}} (-1)^{r-a_{i,j}} q_i^{-r} E_i^{(-a_{i,j}-r)} E_j E_i^{(r)} \quad \text{if } i \neq j$$

$$T_i(F_i) \doteqdot -K_i^{-1} E_i, \quad T_i(F_j) \doteqdot \sum_{r=0}^{-a_{i,j}} (-1)^{r-a_{i,j}} q_i^{r} F_i^{(r)} F_j F_i^{(-a_{i,j}-r)} \quad \text{if } i \neq j$$

$$T_i(K_\beta) \doteqdot K_{s_i(\beta)} \quad \forall \beta \in Q$$

where $E_i^{(r)} \doteqdot \frac{E_i^r}{[r]_{q_i}!}$, $F_i^{(r)} \doteqdot \frac{F_i^r}{[r]_{q_i}!}$.
Moreover $\forall \tau \in \mathcal{T}$ define $T_\tau : \mathcal{U}_q \to \mathcal{U}_q$ to be the $\mathbb{C}(q)$-automorphism of $\mathcal{U}_q$ given by:

$$T_\tau(E_i) \doteqdot E_{\tau(i)}, \quad T_\tau(F_i) \doteqdot F_{\tau(i)}, \quad T_\tau(K_i) \doteqdot K_{\tau(i)} \quad \forall i \in I.$$

Theorem 7.
$\forall i\in I$ $T_i:\mathcal{U}_q\to\mathcal{U}_q$ is well defined and it is a $\mathbb{C}(q)$-automorphism; T_i^{-1} is given by:

$$T_i^{-1}(E_i)=-K_i^{-1}F_i,$$

$$T_i^{-1}(E_j)=\sum_{r=0}^{-a_{i,j}}(-1)^{r-a_{i,j}}q_i^{-r}E_i^{(r)}E_jE_i^{(-a_{i,j}-r)}\quad\text{if } i\neq j$$

$$T_i^{-1}(F_i)=-E_iK_i,$$

$$T_i^{-1}(F_j)=\sum_{r=0}^{-a_{i,j}}(-1)^{r-a_{i,j}}q_i^{r}F_i^{(-a_{i,j}-r)}F_jF_i^{(r)}\quad\text{if } i\neq j$$

$$T_i^{-1}(K_\beta)=K_{s_i(\beta)}\quad\forall\beta\in Q.$$

Analogously $\forall\tau\in\mathcal{T}$ T_τ is a well defined $\mathbb{C}(q)$-automorphism of $\mathcal{U}_q$ and

$$T_\tau^{-1}=T_{\tau^{-1}}.$$

These automorphisms define an action of the extended braid group $\tilde{\mathcal{B}}$ on $\mathcal{U}_q$ such that $T_w^{\pm1}(\mathcal{U}_{q,\alpha})=\mathcal{U}_{q,w^{\pm1}(\alpha)}$ and $T_w^{\pm1}(K_\alpha)=K_{w^{\pm1}(\alpha)}$ $\forall w\in\tilde{W}$ $\forall\alpha\in Q$. In particular $\forall\tau\in\mathcal{T}$ T_τ is an automorphism of $\mathcal{U}_q$ and $T_\tau T_i=T_{\tau(i)}T_\tau$ $\forall i\in I$.

Proof: See Lusztig [18]. □

Remark 8.
$\forall w\in\tilde{W}$ $\forall i\in I$ such that $l(ws_i)=l(w)+1$ we have that $T_w(E_i)\in\mathcal{U}_q^+$ and also $T_{w^{-1}}^{-1}(E_i)\in\mathcal{U}_q^+$. In particular if $l(ws_i)=l(w)+1$ and $w(\alpha_i)=\alpha_j$, then $T_w(E_i)=E_j=T_{w^{-1}}^{-1}(E_i)$.
This in particular implies that if A is of untwisted affine type, $x\in P_{0,+}^{\vee}$ and $\langle x,\alpha_i\rangle=0$, then

$$T_x(E_i)=E_i,\quad T_x(F_i)=F_i,\quad T_x(K_i)=K_i,\quad\text{and}\quad T_xT_i=T_iT_x.$$

Proof: For the proof see Lusztig [15]. □

I end this paragraph defining an important involution of $\mathcal{U}_q$, denoted by Ω, which allows to transfer to $\mathcal{U}_q^-$ the results obtained for $\mathcal{U}_q^+$, and viceversa.

Definition 9.
Ω is the $\mathbb{C}$-antilinear antihomomorphism of $\mathcal{U}_q$ defined by $\Omega(q)\doteqdot q^{-1}$ and, $\forall i\in I$,

$$\Omega(E_i)\doteqdot F_i,\quad\Omega(F_i)\doteqdot E_i,\quad\Omega(K_i)\doteqdot K_i^{-1}.$$

Remark 10.
$\Omega\mathcal{U}_q^0=\mathcal{U}_q^0$ and $\Omega\mathcal{U}_q^+=\mathcal{U}_q^-$; moreover it is immediate to verify that Ω^2=id: in particular Ω is an involution, hence an antiautomorphism; furthermore $T_w\Omega=\Omega T_w$ $\forall w\in\tilde{W}$.

§B3. Hopf algebra structure on $\mathcal{U}_q$.

I recall here how to define a structure of Hopf algebra on $\mathcal{U}_q$, and state some of its properties.

Definition 1.
Let us denote by Δ the homomorphism of $\mathbb{C}(q)$-algebras $\Delta : \mathcal{U}_q \to \mathcal{U}_q \otimes_{\mathbb{C}(q)} \mathcal{U}_q$ defined on the generators by:

$$\Delta(E_i) \doteqdot E_i \otimes 1 + K_i \otimes E_i,$$

$$\Delta(F_i) \doteqdot 1 \otimes F_i + F_i \otimes K_i^{-1},$$

$$\Delta(K_i) \doteqdot K_i \otimes K_i.$$

Denote by ϵ the homomorphism of $\mathbb{C}(q)$-algebras $\epsilon : \mathcal{U}_q \to \mathbb{C}(q)$ given by:

$$\epsilon(E_i) \doteqdot 0, \quad \epsilon(F_i) \doteqdot 0, \quad \epsilon(K_i) \doteqdot 1 \quad \forall i \in I.$$

Define S to be the antiautomorphism of $\mathbb{C}(q)$-algebra $S : \mathcal{U}_q \to \mathcal{U}_q$ given on the generators by:

$$S(E_i) \doteqdot -K_i^{-1}E_i, \quad S(F_i) \doteqdot -F_iK_i, \quad S(K_i) \doteqdot K_i^{-1}.$$

Δ, ϵ and S are well defined.

Theorem 2.
Δ, ϵ and S define a Hopf algebra structure $(\mathcal{U}_q, \Delta, \epsilon, S)$ on $\mathcal{U}_q$, where Δ is the comultiplication, ϵ the counit and S the antipode.
This means that, if $\mathbf{m}:\mathcal{U}_q \otimes_{\mathbb{C}(q)} \mathcal{U}_q \to \mathcal{U}_q$ and $\mathbf{1}:\mathbb{C}(q) \to \mathcal{U}_q$ are respectively the multiplication and the unit of $\mathcal{U}_q$, then
1) $\mathbf{m}$ is $\mathbb{C}(q)$-bilinear and $\mathbf{1}$ is $\mathbb{C}(q)$-linear;
2) $\mathbf{m} \circ (\mathbf{m} \otimes \mathrm{id}) = (\mathbf{m} \otimes \mathrm{id}) \circ \mathbf{m}$ and $\mathbf{m} \circ (\mathrm{id} \otimes \mathbf{1}) = \mathrm{id} = \mathbf{m} \circ (\mathbf{1} \otimes \mathrm{id})$, where one identifies $\mathcal{U}_q \otimes_{\mathbb{C}(q)} \mathbb{C}(q)$ and $\mathbb{C}(q) \otimes_{\mathbb{C}(q)} \mathcal{U}_q$ with $\mathcal{U}_q$;
3) $(\Delta \otimes \mathrm{id}) \circ \Delta = (\mathrm{id} \otimes \Delta) \circ \Delta$ and $(\mathrm{id} \otimes \epsilon) \circ \Delta = \mathrm{id} = \Delta \circ (\mathrm{id} \otimes \epsilon)$;
4) $\mathbf{m} \circ (\mathrm{id} \otimes S) \circ \Delta = \mathrm{id} = \mathbf{m} \circ (S \otimes \mathrm{id}) \circ \Delta$.

Proof: See Lusztig [18]. □

Remark 3.
If $x \in \mathcal{U}_{q,\alpha}$ then $\Delta(x) = \sum_i x_i \otimes x_i'$ where $x_i x_i' \in \mathcal{U}_{q,\alpha}$; if moreover $x \in \mathcal{U}_{q,\alpha}^+$ we have that $x_i \in \mathcal{U}_q^{\geq 0}$ and $x_i' \in \mathcal{U}_q^+$.

Proof: It immediately follows from the definition of Δ on the generators, see Lusztig [18]. □

Definition 4.
σ denotes the automorphism of $\mathbb{C}(q)$-algebra $\sigma : \mathcal{U}_q \otimes_{\mathbb{C}(q)} \mathcal{U}_q \to \mathcal{U}_q \otimes_{\mathbb{C}(q)} \mathcal{U}_q$ given by $\sigma(x \otimes y) \doteqdot y \otimes x$.
Of course σ is well defined.

Lemma 5.
$\forall i \in I\ \forall m \in \mathbb{N}$

$$\Delta(E_i^m) = \sum_{r=0}^{m} q_i^{-r(m-r)} \begin{bmatrix} m \\ r \end{bmatrix}_{q_i} K_i^{m-r} E_i^r \otimes E_i^{m-r},$$

$$\Delta(F_i^m) = \sum_{r=0}^{m} q_i^{r(m-r)} \begin{bmatrix} m \\ r \end{bmatrix}_{q_i} F_i^r \otimes F_i^{m-r} K_i^{-r}.$$

Proof: For the assertion involving E_i I use induction on m, the case $m = 0, 1$ being obvious. Let $m > 1$; then

$$\Delta(E_i^m) = \sum_{r=0}^{m-1} q_i^{-r(m-r-1)} \begin{bmatrix} m-1 \\ r \end{bmatrix}_{q_i} K_i^{m-1-r} E_i^r \otimes E_i^{m-r-1}(E_i \otimes 1 + K_i \otimes E_i) =$$

$$= K_i^m \otimes E_i^m + E_i^m \otimes 1 +$$

$$+ \sum_{r=1}^{m-1} \left(q_i^{-(r-1)(m-r)} \begin{bmatrix} m-1 \\ r-1 \end{bmatrix}_{q_i} + q_i^{-2r-r(m-r-1)} \begin{bmatrix} m-1 \\ r \end{bmatrix}_{q_i} \right) K_i^{m-r} E_i^r \otimes E_i^{m-r}$$

from which the claim follows because

$$q_i^{-(r-1)(m-r)} \begin{bmatrix} m-1 \\ r-1 \end{bmatrix}_{q_i} + q_i^{-2r-r(m-r-1)} \begin{bmatrix} m-1 \\ r \end{bmatrix}_{q_i} =$$

$$= \frac{[m-1]_{q_i}!}{[r]_{q_i}![m-r]_{q_i}!} (q_i^{-(r-1)(m-r)}[r]_{q_i} + q_i^{-2r-r(m-r-1)}[m-r]_{q_i}) =$$

$$= q_i^{-r(m-r)} \begin{bmatrix} m \\ r \end{bmatrix}_{q_i}.$$

On the other hand remark that $\Delta(F_i) = (\Omega \otimes \Omega)\sigma\Delta(E_i)$, so that

$$\Delta(F_i^m) = (\Omega \otimes \Omega)\sigma\Delta(E_i^m),$$

which is the thesis. □

Definition 6.
Let $x \in \mathcal{U}_q$; then ad$x : \mathcal{U}_q \to \mathcal{U}_q$ denotes the $\mathbb{C}(q)$-linear map defined by

$$\mathrm{ad}x(y) \doteqdot \sum_i x_i y S(x_i') \quad \text{where} \quad \Delta(x) = \sum_i x_i \otimes x_i'.$$

Notice that ad$x$ is well defined: indeed
$\mathrm{ad}x(y) = \mathbf{m}(\mathbf{m} \otimes \mathrm{id})(\sigma \otimes \mathrm{id})(\mathrm{id} \otimes \mathrm{id} \otimes S)(\mathrm{id} \otimes \Delta)(y \otimes x)$.

Moreover ad$x$ is a $\mathbb{C}(q)$-linear transformation of $\mathcal{U}_q$. □

Proposition 7.
ad : $\mathcal{U}_q \to \text{End}_{\mathbb{C}(q)}\mathcal{U}_q$ is a representation of $\mathcal{U}_q$, which is called the adjoint representation of $\mathcal{U}_q$.
Proof: ad is obviously $\mathbb{C}(q)$-linear; moreover, if

$$\Delta(x) = \sum_i x_i \otimes x_i' \text{ and } \Delta(y) = \sum_j y_j \otimes y_j',$$

$$\text{ad}xy(z) = \sum_{ij} x_i y_j z S(x_i' y_j') = \sum_i x_i \Big(\sum_j y_j z S(y_j')\Big) S(x_i') = \text{ad}x(\text{ad}y(z)),$$

so that adxy = ad$x$ad$y$. □

Proposition 8.
$\text{ad}E_i^m(E_j) = 0 \ \forall i, j = 1, ..., n$ such that $i \neq j$, $\forall m \geq 1 - a_{i,j}$.
Proof:

$$\text{ad}E_i^{1-a_{i,j}}(E_j) =$$

$$= \sum_{r=0}^{1-a_{i,j}} q_i^{-r(1-a_{i,j}-r)} \begin{bmatrix} 1-a_{i,j} \\ r \end{bmatrix}_{q_i} K_i^{1-a_{i,j}-r} E_i^r E_j S(E_i^{1-a_{i,j}-r}) =$$

$$= (-1)^{1-a_{i,j}} \sum_{r=0}^{1-a_{i,j}} (-1)^r \begin{bmatrix} 1-a_{i,j} \\ r \end{bmatrix}_{q_i} E_i^r E_j E_i^{1-a_{i,j}-r} = 0.$$

Hence if $m \geq 1 - a_{i,j}$

$$\text{ad}E_i^m(E_j) = \text{ad}E_i^{m-1+a_{i,j}} \text{ad}E_i^{1-a_{i,j}}(E_j) = 0.$$

□

Theorem 9.
There exists a unique $\mathbb{C}(q)$-bilinear form

$$(\cdot,\cdot) : \mathcal{U}_q^{\geq 0} \times \mathcal{U}_q^{\leq 0} \to \mathbb{C}(q)$$

such that
1) $(x, y_1 y_2) = (\Delta(x), y_1 \otimes y_2) \ \forall x \in \mathcal{U}_q^{\geq 0}, y_1, y_2 \in \mathcal{U}_q^{\leq 0}$;
2) $(x_1 x_2, y) = (x_2 \otimes x_1, \Delta(y)) \ \forall x_1, x_2 \in \mathcal{U}_q^{\geq 0}, y \in \mathcal{U}_q^{\leq 0}$;
3) $(K_\lambda, K_\mu) = q^{-(\lambda|\mu)} \ \forall \lambda, \mu \in Q$;
4) $(K_\lambda, F_i) = 0 = (E_i, K_\lambda) \ \forall i \in I, \lambda \in Q$;
5) $(E_i, F_j) = \dfrac{\delta_{ij}}{q_i^{-1} - q_i} \ \forall i, j \in I$.

Proof: See Tanisaki [23]. □

Remark 10.
$(x,y)=0$ if $x\in\mathcal{U}_{q,\eta}$ and $y\in\mathcal{U}_{q,-\tilde\eta}$ with $\eta\neq\tilde\eta$.
Moreover $(\cdot,\cdot)|_{\mathcal{U}_{q,\eta}\times\mathcal{U}_{q,-\eta}}$ is non degenerate.

Proof: For the proof see Tanisaki [23]. □

Part C. Bases of type Poincaré-Birkhoff-Witt for $\mathcal{U}_q(\hat{\mathfrak{g}})$ (untwisted affine case).

§C1. Bases associated with strictly dominant weights.

For the results of this and the following paragraphs see Beck [2].
Let us consider the following notations:

Definition 1.
$\forall\underline{i}\doteqdot(i_r)\in I^{\mathbb{Z}}$, $\forall m\in\mathbb{Z}$ let us define

$$\beta(m,\underline{i})\doteqdot\begin{cases} s_{i_1}\cdot\ldots\cdot s_{i_{m-1}}(\alpha_{i_m}), & \text{if } m>0\\ s_{i_0}s_{i_{-1}}\cdot\ldots\cdot s_{i_{m+1}}(\alpha_{i_m}) & \text{if } m\leq 0,\end{cases}$$

$$E(m,\underline{i})\doteqdot\begin{cases} T_{i_1}\cdot\ldots\cdot T_{i_{m-1}}(E_{i_m}) & \text{if } m>0\\ T_{i_0}^{-1}T_{i_{-1}}^{-1}\cdot\ldots\cdot T_{i_{m+1}}^{-1}(E_{i_m}) & \text{if } m\leq 0.\end{cases}$$

Remark 2.
$\beta(m,i)\in R^{\mathrm{re}}$ and $E(m,\underline{i})\in\mathcal{U}_{q,\beta(m,\underline{i})}$ $\forall m\in\mathbb{Z}$, $\forall\underline{i}\in I^{\mathbb{Z}}$, but it's not generally true that $E(m,\underline{i})\in\mathcal{U}_q^+$, neither that $\beta(m,\underline{i})\in R_+^{\mathrm{re}}$.
The first step in the search for a basis of type PBW of $\mathcal{U}_q(\hat{\mathfrak{g}})$ is to find conditions on $\underline{i}$ so that $E(m,\underline{i})\in\mathcal{U}_q^+$ $\forall m\in\mathbb{Z}$ (and in particular $\beta(m,\underline{i})\in R_+^{\mathrm{re}}$).

Definition 3.
Let $w\in W$ and $w=s_{i_1}\cdot\ldots\cdot s_{i_N}$ be a reduced expression of w; define $\underline{i}(w)\in I^{\mathbb{Z}}$ by $i_r(w)\doteqdot i_s$ where $0<s\leq N$ and $r\equiv s \pmod N$.
Also set $\beta_w(m)\doteqdot\beta(m,\underline{i}(w))$ and $E_w(m)\doteqdot E(m,\underline{i}(w))$.
Remark that both $\beta_w(m)$ and $E_w(m)$ depend on the reduced expression of w and not only on w; anyway I generally omit the dependence on $\underline{i}(w)$ (which will be usually fixed).

Proposition 4.
Let $x\in P_{0,+}^{\check{}}\cap W=P_{0,+}^{\check{}}\cap Q_0^{\check{}}$ with $l(x)=N$; then
i) $\forall a,b\in\mathbb{Z}$ with $a\leq b$ $s_{i_a(x)}\cdot\ldots\cdot s_{i_b(x)}$ is a reduced expression;
ii) $E_x(m)\in\mathcal{U}_q^+$ $\forall m\in\mathbb{Z}$;
iii) $\beta_x(m)\in R_+$ $\forall m\in\mathbb{Z}$;
iv) $\beta_x(r)=\beta_x(s)$ if and only if $r=s$;
v) let $m>0$; then

$$\{\beta_x(r)|1\leq r\leq mN\}=\{-\alpha+r\delta|\alpha\in\mathring{R}_{0,+},0<r\leq m\langle x,\alpha\rangle\}$$

and

$$\{\beta_x(r)| - mN < r \leq 0\} = \{\alpha + r\delta | \alpha \in R_{0,+}, 0 \leq r < m\langle x, \alpha\rangle\};$$

vi) $\{\beta_x(r)|r \geq 1\} = \{-\alpha + m\delta | m > 0, \alpha \in R_{0,+}, \langle x, \alpha\rangle > 0\}$ and $\{\beta_x(r)|r \leq 0\} = \{\alpha + m\delta | m \geq 0, \alpha \in R_{0,+}, \langle x, \alpha\rangle > 0\}$.

Proof: i) Since $x \in P^{\vee}_{0,+}$, $l(mx) = ml(x)$ $\forall m \in \mathbb{N}$, so that $s_{i_1(x)} \cdot \ldots \cdot s_{i_{mN}(x)}$ is a reduced expression $\forall m \in \mathbb{N}$; in particular $s_{i_a(x)} \cdot \ldots \cdot s_{i_b(x)}$ is a reduced expression $\forall a, b \in \mathbb{Z}$ such that $1 \leq a \leq b$; the claim follows remarking that $\underline{i}$ is a cyclic sequence.
ii) immediately follows from i) and from remark 1.B2.8.
iii) follows from ii) and from the fact that $E_x(m) \in \mathcal{U}_{q,\beta_x(m)}$ with $\beta_x(m) \in R^{\text{re}}$.
iv) Suppose $\beta_x(a) = \beta_x(b)$ with $a < b$; then,

$$\text{if } a < b \leq 0 \text{ or } 1 \leq a < b, \quad s_{i_{a+1}(x)} \cdot \ldots \cdot s_{i_{b-1}(x)}(\alpha_{i_b(x)}) = -\alpha_{i_a(x)} < 0,$$

so that $s_{i_{a+1}(x)} \cdot \ldots \cdot s_{i_b(x)}$ is not reduced (contradiction), while,

$$\text{if } a \leq 0 < b, \quad s_{i_a(x)} \cdot \ldots \cdot s_{i_{b-1}(x)}(\alpha_{i_b(x)}) = -\alpha_{i_a(x)} < 0,$$

so that $s_{i_a(x)} \cdot \ldots \cdot s_{i_b(x)}$ is not reduced, which again contradicts i).
v) By proposition 1.A1.17 $\{\beta_x(r)|1 \leq r \leq mN\} = \{\alpha \in R_+ | x^{-m}(\alpha) < 0\}$; but $\alpha \in R_+ \Rightarrow \alpha = \beta + r\delta$ with $\beta \in R_{0,+}$ and $r \geq 0$, or $\beta \in R_{0,-}$ and $r > 0$; on the other hand $x^{-m}(\alpha) = x^{-m}(\beta) + r\delta = \beta + (\langle mx, \beta\rangle + r)\delta$, so that necessarily $\langle mx, \beta\rangle < 0$; but $x \in P^{\vee}_{0,+}$ implies $\langle x, \alpha_i\rangle \geq 0$ $\forall i \in I$, hence $\beta \in -R_{0,+}$ and $\langle mx, \beta\rangle + r \leq 0$; then

$$\{\beta_x(r)|1 \leq r \leq mN\} = \{-\alpha + r\delta | \alpha \in R_{0,+}, 0 < r \leq m\langle x, \alpha\rangle\}.$$

Analogously $\{\beta_x(r)| - mN < r \leq 0\} = \{\alpha \in R_+ | x^m(\alpha) < 0\}$; but

$$x^m(\beta + r\delta) = x^m(\beta) + r\delta = \beta + (r - \langle mx, \beta\rangle)\delta,$$

so that necessarily $\langle mx, \beta\rangle > 0$; hence $\beta \in R_{0,+}$ and $r - \langle mx, \beta\rangle < 0$; then

$$\{\beta_x(r)| - mN < r \leq 0\} = \{\alpha + r\delta | \alpha \in R_{0,+}, 0 \leq r < m\langle x, \alpha\rangle\}.$$

vi) follows immediately from v). □

Corollary 5.
If $x \in P^{\vee}_{0,++} \cap W$ then

$$\{\beta_x(r)|r \geq 1\} = \{-\alpha + m\delta | \alpha \in R_{0,+}, m > 0\} \quad \text{and}$$

$$\{\beta_x(r)|r \leq 0\} = \{\alpha + m\delta | \alpha \in R_{0,+}, m \geq 0\}.$$

In particular $r \mapsto \beta_x(r)$ is a bijection from $\mathbb{Z}$ onto R_+^{re}.

Proof: It follows immediately from the fact that $x \in P_{0,++}^{\check{}}$ implies $\langle x, \alpha\rangle > 0$ $\forall \alpha > 0$.

Definition 6.
Denote by $\mathcal{U}_{q,\mathrm{im}}^+(\underline{i})$ the $\mathbb{C}(q)$-subalgebra of $\mathcal{U}_q^+$ consisting of the elements $y \in \mathcal{U}_q^+$ such that $T_{i_m}^{-1} \cdot \ldots \cdot T_{i_1}^{-1}(y) \in \mathcal{U}_q^+$, $T_{i_{-m}} T_{i_{-m+1}} \cdot \ldots \cdot T_{i_0}(y) \in \mathcal{U}_q^+$ $\forall m \in \mathbb{N}$ and, for $w \in W$, set $\mathcal{U}_{q,\mathrm{im}}^+(w) \doteqdot \mathcal{U}_{q,\mathrm{im}}^+(\underline{i}(w))$.

Definition 7.
Let $\underline{i} \in I^{\mathbb{Z}}$ and let B be a basis of $\mathcal{U}_{q,\mathrm{im}}^+(\underline{i})$. $\forall b \in B$, $\forall \underline{c} \in \oplus_{\mathbb{Z}} \mathbb{N}$ let $L^{(\underline{i})}(\underline{c}, b)$ be the element of $\mathcal{U}_q$ defined by

$$L^{(\underline{i})}(\underline{c}, b) \doteqdot E(1,\underline{i})^{c_1} E(2,\underline{i})^{c_2} \cdot \ldots \cdot E(r,\underline{i})^{c_r} b E(s,\underline{i})^{c_s} E(s+1,\underline{i})^{c_{s+1}} \cdot \ldots \cdot E(0,\underline{i})^{c_0},$$

where $r > 0$, $s \leq 0$ are such that $c_t \neq 0 \Rightarrow s \leq t \leq r$.
If $w \in W$, $L^{(w)}(\underline{c}, b) \doteqdot L^{(\underline{i}(w))}(\underline{c}, b)$, that is

$$L^{(w)}(\underline{c}, b) = E_w(1)^{c_1} E_w(2)^{c_2} \cdot \ldots \cdot E_w(r)^{c_r} b E_w(s)^{c_s} E_w(s+1)^{c_{s+1}} \cdot \ldots \cdot E_w(0)^{c_0}.$$

Of course if $E_w(r) \in \mathcal{U}_q^+$ for all $r \in \mathbb{Z}$, then also

$$L^{(w)}(\underline{c}, b) \in \mathcal{U}_q^+ \quad \forall b \in B, \quad \forall \underline{c} \in \oplus_{\mathbb{Z}} \mathbb{N}.$$

Then we have the following theorem

Theorem 8.
$\forall x \in P_{0,++}^{\check{}} \cap Q_0^{\check{}}$, for any reduced expression $s_{i_1} \cdot \ldots \cdot s_{i_N}$ of x, and $\forall B$ $\mathbb{C}(q)$-basis of $\mathcal{U}_{q,\mathrm{im}}^+(x)$, the set

$$\{L^{(x)}(\underline{c}, b) | \underline{c} \in \oplus_{\mathbb{Z}} \mathbb{N}, b \in B\}$$

is a basis of $\mathcal{U}_q^+$.
More precisely

$$\mathcal{U}_{q,\mathrm{im}}^+(x) = \mathbb{C}(q)[\tilde{E}_{(m\delta,i)} | i \in I_0, m > 0]$$

is independent of $x \in P_{0,++}^{\check{}}$

$$(\text{where } \tilde{E}_{(m\delta,i)} \doteqdot T_{\omega_i^{\check{}}}^m (K_i^{-1} F_i) E_i - q_i^{-2} E_i T_{\omega_i^{\check{}}}^m (K_i^{-1} F_i) \in \mathcal{U}_{q,m\delta}^+),$$

so that

$$\{ \prod_{m>0, i \in I_0} \tilde{E}_{(m\delta,i)}^{r_{m,i}} | \underline{r} \in \oplus_{(\mathbb{N}\setminus\{0\}) \times I_0} \mathbb{N}\}$$

is a basis of $\mathcal{U}_{q,\mathrm{im}}^+ \doteqdot \mathcal{U}_{q,\mathrm{im}}^+(x)$, which is a commutative subalgebra of $\mathcal{U}_q^+$.

Proof: See Beck [2]. □

Remark 9.
$T_{\omega_i^\vee}|_{\mathcal{U}^+_{q,\mathrm{im}}} = \mathrm{id}_{\mathcal{U}^+_{q,\mathrm{im}}}\ \forall i\in I_0$.
I want now to extend slightly the result of the previous theorem.

Proposition 10.
Let $\underline{i} \doteqdot (i_r)\in I^{\mathbb{Z}}$ and $\{x_{m,i}\}_{m>0,i\in I_0}\subset \cup_{\alpha\in R_+^{\mathrm{im}}}\mathcal{U}^+_{q,\alpha}$ be such that

1) the subalgebra $\tilde{\mathcal{U}}^+_{q,\mathrm{im}}$ of $\mathcal{U}_q^+$ generated by $\{x_{m,i}\}_{m>0,i\in I_0}$ is a commutative algebra isomorphic to the algebra of polynomials $\mathbb{C}[X_{m,i}|m>0,i\in I_0]\hookrightarrow \mathcal{U}_q^+$ via the map induced by $X_{m,i}\mapsto x_{m,i}$;
2) $T_{i_m}^{-1}\cdot\ldots\cdot T_{i_1}^{-1}(\tilde{\mathcal{U}}^+_{q,\mathrm{im}})\subset\mathcal{U}_q^+$, $T_{i_{-m}}T_{i_{-m+1}}\cdot\ldots\cdot T_{i_0}(\tilde{\mathcal{U}}^+_{q,\mathrm{im}})\subset\mathcal{U}_q^+\ \forall m\in\mathbb{N}$;
3) $\{L(\underline{c},\underline{r},\underline{i})|\underline{c}\in\oplus_{\mathbb{Z}}\mathbb{N},\underline{r}\in\oplus_{(\mathbb{N}\setminus\{0\})\times I_0}\mathbb{N}\}$, where

$$L(\underline{c},\underline{r},\underline{i})\doteqdot E(1,\underline{i})^{c_1}E(2,\underline{i})^{c_2}\cdot\ldots\cdot E(s,\underline{i})^{c_s}\cdot$$

$$\cdot\Big(\prod_{(m,i)\in(\mathbb{N}\setminus\{0\})\times I} x_{m,i}^{r_{m,i}}\Big)E(t,\underline{i})^{c_t}E(t+1,\underline{i})^{c_{t+1}}\cdot\ldots\cdot E(0,\underline{i})^{c_0}$$

is a basis of $\mathcal{U}_q^+$.
Then $\forall a\in\mathbb{Z}$ if we define $i_r^{(a)}\doteqdot i_{r+a}$,

$$x_{m,i}^{(a)}\doteqdot\begin{cases} T_{i_a}^{-1}\cdot\ldots\cdot T_{i_1}^{-1}(x_{m,i}) & \text{if } a\geq 0\\ T_{i_{a+1}}\cdot\ldots\cdot T_{i_0}(x_{m,i}) & \text{if } a<0,\end{cases}$$

$(i_r^{(a)})\in I^{\mathbb{Z}}$ and $\{x_{m,i}^{(a)}\}_{m>0,i\in I}\subset\mathcal{U}_q^+$ satisfy conditions 1), 2) and 3).

Proof: The thesis is obvious if $a=0$; moreover, of course, $\underline{i}^{(a+b)}=(\underline{i}^{(a)})^{(b)}$, $\underline{x}^{(a+b)}=(\underline{x}^{(a)})^{(b)}\ \forall a,b\in\mathbb{Z}$; hence it is enough to prove the thesis for $a=\pm1$. Conditions 1) and 2) are immediate. Notice that

$$E(m,\underline{i}^{(1)})=\begin{cases} T_{i_1}^{-1}(E(m+1,\underline{i})) & \text{if } m\neq 0\\ -T_{i_1}^{-1}(F_{i_1}K_{i_1}) & \text{if } m=0\end{cases}$$

and similarly

$$E(m,\underline{i}^{(-1)})=\begin{cases} T_{i_0}(E(m-1,\underline{i})) & \text{if } m\neq 1\\ -T_{i_0}(K_{i_0}^{-1}F_{i_0}) & \text{if } m=1.\end{cases}$$

Now recall that $\mathcal{U}_q\cong\mathcal{U}_q^-\otimes\mathcal{U}_q^0\otimes\mathcal{U}_q^+\cong\mathcal{U}_q^+\otimes\mathcal{U}_q^0\otimes\mathcal{U}_q^-$, that

$$\mathcal{U}_q^0\cong\mathbb{C}(q)[K_i^{\pm1}|i\in I]=\oplus_{\lambda\in Q}\mathbb{C}(q)K_\lambda,\quad \mathcal{U}_q^-=\Omega(\mathcal{U}_q^+)$$

and that $T_i\in\mathrm{Aut}\mathcal{U}_q\ \forall i$.

For $\lambda \in Q$, $\underline{c}, \underline{c}' \in \oplus_{\mathbb{Z}}\mathbb{N}$, $\underline{r}, \underline{r}' \in \oplus_{(\mathbb{N}\setminus\{0\})\times I_0}\mathbb{N}$, define $\tilde{L}(\underline{c}, \underline{c}', \underline{r}, \underline{r}', \lambda, \underline{i})$ and $\hat{L}(\underline{c}, \underline{c}', \underline{r}, \underline{r}', \lambda, \underline{i})$ by

$$\tilde{L}(\underline{c}, \underline{c}', \underline{r}, \underline{r}', \lambda, \underline{i}) \doteqdot \Omega(L(\underline{c}', \underline{r}', \underline{i}))K_\lambda L(\underline{c}, \underline{r}, \underline{i})),$$

$$\hat{L}(\underline{c}, \underline{c}', \underline{r}, \underline{r}', \lambda, \underline{i}) \doteqdot L(\underline{c}, \underline{r}, \underline{i})K_\lambda \Omega(L(\underline{c}', \underline{r}', \underline{i})).$$

Then $\{\tilde{L}(\underline{c}, \underline{c}', \underline{r}, \underline{r}', \lambda, \underline{i})\}$ and $\{\hat{L}(\underline{c}, \underline{c}', \underline{r}, \underline{r}', \lambda, \underline{i})\}$ are bases of $\mathcal{U}_q$, so also

$$B_1 \doteqdot \{T_{i_1}^{-1}(\hat{L}(\underline{c}, \underline{c}', \underline{r}, \underline{r}', \lambda, \underline{i}))\} \text{ and } B_{-1} \doteqdot \{T_{i_0}(\tilde{L}(\underline{c}, \underline{c}', \underline{r}, \underline{r}', \lambda, \underline{i}))\}$$

are bases of $\mathcal{U}_q$. In particular any subset of B_1 is $\mathbb{C}(q)$-linearly independent in $\mathcal{U}_q$ and so is any subset of B_{-1}. But

$$L(\underline{c}, \underline{r}, \underline{i}^{(1)}) = L(\underline{c}', \underline{r}, \underline{i}^{(1)})E(0, \underline{i}^{(1)})^{c_0}$$

where

$$c'_m \doteqdot \begin{cases} c_m & \text{if } m \neq 0 \\ 0 & \text{if } m = 0, \end{cases}$$

so that

$$L(\underline{c}, \underline{r}, \underline{i}^{(1)}) = (-1)^{c_0} q_{i_1}^{c_0(c_0+1)} T_{i_1}^{-1}(L(\underline{\hat{c}}, \underline{r}, \underline{i})K_{i_1}^{c_0} F_{i_1}^{c_0})$$

where

$$\hat{c}_m \doteqdot c'_{m-1} = \begin{cases} c_{m-1} & \text{if } m \neq 1 \\ 0 & \text{if } m = 1, \end{cases}$$

hence

$$L(\underline{c}, \underline{r}, \underline{i}^{(1)}) = (-1)^{c_0} q_{i_1}^{c_0(c_0+1)} T_{i_1}^{-1}(\hat{L}(\underline{\hat{c}}, c_0\underline{\delta}^{(1)}, \underline{r}, \underline{0}, c_0\alpha_{i_1}, \underline{i}))$$

where $\delta_m^{(1)} \doteqdot \delta_{1,m}$.

From this it is clear that $\{L(\underline{c}, \underline{r}, \underline{i}^{(1)}) | \underline{c} \in \oplus_{\mathbb{Z}}\mathbb{N}, \underline{r} \in \oplus_{(\mathbb{N}\setminus\{0\})\times I_0}\mathbb{N}\}$ is linearly independent in $\mathcal{U}_q^+$, since $(\underline{\hat{c}}, c_0\underline{\delta}^{(1)}, \underline{r}) = (\underline{\hat{d}}, d_0\underline{\delta}^{(1)}, \underline{s}) \Rightarrow \underline{c} = \underline{d}$ and $\underline{r} = \underline{s}$. Similarly one sees that

$$L(\underline{c}, \underline{r}, \underline{i}^{(-1)}) = (-1)^{c_1} q_{i_0}^{c_1(c_1+1)} T_{i_0}(\tilde{L}(\underline{\tilde{c}}, c_1\underline{\delta}^{(0)}, \underline{r}, \underline{0}, -c_1\alpha_{i_0}, \underline{i}))$$

where

$$\tilde{c}_m \doteqdot \begin{cases} c_{m+1} & \text{if } m \neq 0 \\ 0 & \text{if } m = 0, \end{cases}$$

and $\delta_m^0 \doteqdot \delta_{0,m}$.

Hence also $\{L(\underline{c}, \underline{r}, \underline{i}^{(-1)}) | \underline{c} \in \oplus_{\mathbb{Z}}\mathbb{N}, \underline{r} \in \oplus_{(\mathbb{N}\setminus\{0\})\times I_0}\mathbb{N}\}$ is linearly independent in $\mathcal{U}_q^+$.

It is then enough to prove that $\forall\eta\in Q$

$$\dim_{\mathbb{C}(q)}(\langle L(\underline{c},\underline{r},\underline{i}^{(1)})\rangle_{\mathbb{C}(q)})\cap\mathcal{U}_{q,\eta})=\dim_{\mathbb{C}(q)}\mathcal{U}^+_{q,\eta}=$$

$$=\dim_{\mathbb{C}(q)}(\langle L(\underline{c},\underline{r},\underline{i}^{(-1)})\rangle_{\mathbb{C}(q)})\cap\mathcal{U}_{q,\eta}).$$

But this follows immediately from the fact that $\dim_{\mathbb{C}(q)}\mathcal{U}^+_{q,\eta}=\mathrm{Par}(\eta)$, that

$$\beta(m,\underline{i}^{(1)})=\begin{cases} s_{i_1}(\beta(m+1,\underline{i})) & \text{if } m\neq 0\\ \alpha_{i_1}=-s_{i_1}(\beta(1,\underline{i})) & \text{if } m=0,\end{cases}$$

$$\beta(m,\underline{i}^{(-1)})=\begin{cases} s_{i_0}(\beta(m-1,\underline{i})) & \text{if } m\neq 1\\ \alpha_{i_0}=-s_{i_0}(\beta(0,\underline{i})) & \text{if } m=1,\end{cases}$$

$s_i(\alpha)=\alpha\ \forall\alpha\in R^{\mathrm{im}}$, and that $s_i(R^{\mathrm{re}}_+\setminus\{\alpha_i\})=R^{\mathrm{re}}_+\setminus\{\alpha_i\}$. □

Thus the following result is proved:

Theorem 11.
Let $x\in P_{\check{0},++}\cap Q_{\check{0}}$ and let $s_{i_1}\cdot\ldots\cdot s_{i_N}$ be a reduced expression of x; $\forall a\in\mathbb{Z}$ let $x^{(a)}\doteqdot s_{i_{a+1}}\cdot\ldots\cdot s_{i_{a+N}}$ where $\underline{i}=\underline{i}(x)$ is as defined in definition 1.C1.3. Then $\forall a\in\mathbb{Z}$

$$l(x^{(a)})=N=l(x)$$

$$l(rx^{(a)})=|r|l(x^{(a)})\ \ \forall r\in\mathbb{Z}$$

$$\{\beta_{x^{(a)}}(r)=\beta(r,\underline{i}^{(a)})|r\in\mathbb{Z}\}=R^{\mathrm{re}}_+$$

$$\text{and}\ \ \exists\{x^{(a)}_{m,i}\}_{m>0,i\in I_0}\subset\oplus_{\alpha\in R^{\mathrm{im}}_+}\mathcal{U}^+_{q,\alpha}\ \ (\text{with } x^{(a)}_{m,i}\in\mathcal{U}^+_{q,m\delta})$$

such that the ordered monomials in $\{E(m,\underline{i}^{(a)}),x^{(a)}_{r,j}|m\in\mathbb{Z},r>0,j\in I_0\}$ form a basis of $\mathcal{U}^+_q$ and $T^{-1}_{i_{a+b}}\cdot\ldots\cdot T^{-1}_{i_{a+1}}(x^{(a)}_{m,i})$, $T_{i_{a-b-1}}\cdot\ldots\cdot T_{i_a}(x^{(a)}_{m,i})\in\mathcal{U}^+_q\ \ \forall b\in\mathbb{N}$.
□

Corollary 12.
Let $x\in P_{\check{0},++}$ and $\underline{i}\doteqdot\underline{i}(x)$; then

$$T_{i_1}\cdot\ldots\cdot T_{i_k}(E)=\sum ef_\gamma K_\gamma\ \ \forall k>0,\forall E\in\mathcal{U}^+_q$$

and

$$T^{-1}_{i_0}\cdot\ldots\cdot T^{-1}_{i_h}(F)=\sum fK_\gamma e_\gamma\ \ \forall h\leq 0,\forall F\in\mathcal{U}^-_q,$$

where $e\in\mathcal{U}^+_q$, $f\in\mathcal{U}^-_q$, $e_\gamma\in\mathcal{U}^+_{q,\gamma}$, $f_\gamma\in\mathcal{U}^-_{q,-\gamma}$.

Proof: Since every $E\in\mathcal{U}^+_q$ (resp. $F\in\mathcal{U}^-_q$) is a linear combination of terms of the form

$$L^{(x^{(k)})}(\underline{c},b)\ \ (\text{resp. } \Omega(L^{(x^{(h-1)})}(\underline{c},b))\ )$$

(see theorem 1.C1.11), I can suppose

$$E = L^{(x^{(k)})}(\underline{c}, b) \text{ and } F = \Omega(L^{(x^{(h-1)})}(\underline{c}, b)).$$

Now

$$T_{i_1} \cdot ...T_{i_k}(b) \in \mathcal{U}_q^+ \quad \text{and}$$

$$T_{i_1} \cdot ...T_{i_k}(E_{x^{(k)}}(r)) \in \mathcal{U}_q^+ \ \forall r > 0 \text{ and } \forall r \leq -k$$

while if $-k < r \leq 0$

$$T_{i_1} \cdot ...T_{i_k}(E_{x^{(k)}}(r)) = T_{i_1} \cdot ...T_{i_k} T_{i_k}^{-1} \cdot ...T_{i_{r+k+1}}^{-1}(E_{i_{r+k}}) =$$

$$= T_{i_1} \cdot ...T_{i_{r+k}}(E_{i_{r+k}}) = -T_{i_1} \cdot ...T_{i_{r+k-1}}(F_{i_{r+k}} K_{i_{r+k}}) = -F_{\beta_{r+k}} K_{\beta_{r+k}}.$$

Thus $T_{i_1} \cdot ...T_{i_k}(L^{(x^{(k)})}(\underline{c}, b) = e f_\gamma K_\gamma$ with $e \in \mathcal{U}_q^+$ and $f_\gamma \in \mathcal{U}_{q,-\gamma}^-$.
Analogously

$$\Omega T_{i_0}^{-1} \cdot ...T_{i_h}^{-1}(b) \in \mathcal{U}_q^- \quad \text{and}$$

$$T_{i_0}^{-1} \cdot ...T_{i_h}^{-1} \Omega(E_{x^{(h-1)}}(r)) \in \mathcal{U}_q^- \ \forall r \leq 0 \text{ and } \forall r > -h+1$$

while if $0 < r \leq 1-h$

$$T_{i_0}^{-1} \cdot ...T_{i_h}^{-1} \Omega(E_{x^{(h-1)}}(r)) = T_{i_0}^{-1} \cdot ...T_{i_h}^{-1} T_{i_h} \cdot ...T_{i_{r+h-2}}(F_{i_{r+h-1}}) =$$

$$= T_{i_0}^{-1} \cdot ...T_{i_{r+h-1}}^{-1}(F_{i_{r+h-1}}) =$$

$$= -T_{i_0}^{-1} \cdot ...T_{i_{r+h}}^{-1}(E_{i_{r+h-1}} K_{i_{r+h-1}}) = -E_{\beta_{r+h-1}} K_{\beta_{r+h-1}},$$

so that $T_{i_0}^{-1} \cdot ...T_{i_h}^{-1} \Omega(L^{(x^{(k)})}(\underline{c}, b) = f K_\gamma e_\gamma$ with $f \in \mathcal{U}_q^-$ and $e_\gamma \in \mathcal{U}_{q,\gamma}^+$. □

§C2. A basis of type PBW for $\mathcal{U}_q(\hat{\mathfrak{g}})$.

Let $\rho^\vee \doteq \frac{1}{2} \sum_{\alpha^\vee \in R_{0,+}^\vee} \alpha^\vee$.
Since $\rho^\vee = \sum_{i=1}^n \omega_i^\vee \in P_{0,++}^\vee$ we have that $2\rho^\vee \in P_{0,++}^\vee \cap Q_0^\vee$.
Then if $x = 2\rho^\vee$ we can associate to any reduced expression of x a basis of type PBW of $\mathcal{U}_q$ (see Beck [2]). I choose a reduced expression $x = s_{i_1} \cdot ... \cdot s_{i_N}$ of x in the following way: $s_{i_1} \cdot ... \cdot s_{i_N}$ is the juxtaposition of reduced expressions of $\tau_0 \omega_1^\vee \tau_1^{-1}$, $\tau_1 \omega_2^\vee \tau_2^{-1}$,..., $\tau_{n-1} \omega_n^\vee \tau_n^{-1}$, $\tau_n \omega_1^\vee \tau_{n+1}^{-1}$,..., $\tau_{2n-1} \omega_n^\vee \tau_{2n}^{-1}$, where $\tau_r \in \mathcal{T}$ is defined by:
$\tau_0 \doteq \mathrm{id}$, τ_r is such that $\tau_{r-1} \omega_r^\vee \tau_r^{-1} \in W$ if $0 < r \leq n$ and $\tau_{r-1} \omega_{r-n}^\vee \tau_r^{-1} \in W$ if $r > n$.
Notice that $\tau_{n+r} = \tau_n \tau_r$ $\forall r \leq n$, $\tau_n^2 = \tau_{2n} = \mathrm{id}$ (because $2\rho^\vee \in Q_0^\vee \subseteq W$) and $\rho^\vee = \left(\prod_{i=1}^n \tau_{i-1} \omega_i^\vee \tau_i^{-1} \right) \tau_n$.
Consequently one can suppose $i_{r+\frac{N}{2}} = \tau_n(i_r)$ $\forall r \leq \frac{N}{2}$.

Remark 1.
The choice of the reduced expression of $2\rho\check{}$ implies that $\forall m\in\mathbb{N}$
$s_{i_1}\cdot\ldots\cdot s_{i_{m+\frac{N}{2}}} = (\rho\check{}\tau_n^{-1})s_{i_{1+\frac{N}{2}}}\cdot\ldots\cdot s_{i_{m+\frac{N}{2}}} = \rho\check{} s_{i_1}\cdot\ldots\cdot s_{i_m}\tau_n^{-1}$ and
$s_{i_0}\cdot\ldots\cdot s_{i_{-m-\frac{N}{2}}} = ((\rho\check{})^{-1}\tau_n)s_{i_{-\frac{N}{2}}}\cdot\ldots\cdot s_{i_{-m-\frac{N}{2}}} = (\rho\check{})^{-1}s_{i_0}\cdot\ldots\cdot s_{i_{-m}}\tau_n$.

Definition 2.
$\forall m\in\mathbb{Z}\ \beta_m \doteqdot \beta_{2\rho\check{}}(m)$.

Definition 3.
$\forall\beta\in\tilde{R}_+$ I define root vectors E_β, $\tilde{E}_\beta$, F_β and $\tilde{F}_\beta$ in the following way:

$$E_\beta = \tilde{E}_\beta \doteqdot E_{2\rho\check{}}(m) = \begin{cases} T_{i_1}\cdot\ldots\cdot T_{i_{m-1}}(E_{i_m}) & \text{if } m>0 \\ T_{i_0}^{-1}\cdot\ldots\cdot T_{i_{m+1}}^{-1}(E_{i_m}) & \text{if } m\leq 0 \end{cases}$$

$$F_\beta = \tilde{F}_\beta \doteqdot \Omega(E_\beta)$$

if $\beta = \beta_{2\rho\check{}}(m)$ for some $m\in\mathbb{Z}$;

$$\tilde{E}_{(m\delta,i)} = T_{\omega_i}^m(K_i^{-1}F_i)E_i - q_i^{-2}E_iT_{\omega_i}^m(K_i^{-1}F_i)$$

$$\tilde{F}_{(m\delta,i)} \doteqdot \Omega(\tilde{E}_{(m\delta,i)})\quad \forall m>0, i\in I_0;$$

$$1-(q_i-q_i^{-1})\sum_{k>0}\tilde{E}_{(k\delta,i)}u^k = \exp\Big((q_i-q_i^{-1})\sum_{k>0}E_{(k\delta,i)}u^k\Big)$$

$$\text{(that is } (q_i-q_i^{-1})\sum_{k>0}E_{(k\delta,i)}u^k = \lg\Big(1-(q_i-q_i^{-1})\sum_{k>0}\tilde{E}_{(k\delta,i)}u^k\Big)\Big)\quad \forall i\in I_0,$$

$$F_{(k\delta,i)} \doteqdot \Omega(E_{(k\delta,i)})\quad \forall k>0, \forall i\in I_0.$$

Proposition 4.
The $\mathbb{C}(q)$-subalgebra of $\mathcal{U}_q$ generated by $\{E_{(m\delta,i)}|m>0, i\in I_0\}$ is the same as the $\mathbb{C}(q)$-subalgebra of $\mathcal{U}_q$ generated by $\{\tilde{E}_{(m\delta,i)}|m>0, i\in I_0\}$ and both are commutative algebras of polynomials (in $\{E_{(m\delta,i)}|m>0, i\in I_0\}$ and $\{\tilde{E}_{(m\delta,i)}|m>0, i\in I_0\}$ respectively).
More precisely, $\forall M>0$, $\forall i\in I_0$ the $\mathbb{Q}[q,q^{-1}]$-linear span of

$$\Big\{\prod_{m_1,\ldots,m_r>0}E_{(m_j\delta,i)}\Big|\sum_{j=1}^r m_j = M\Big\}$$

is the same as the $\mathbb{Q}[q,q^{-1}]$-linear span of

$$\Big\{\prod_{m_1,\ldots,m_r>0}\tilde{E}_{(m_j\delta,i)}\Big|\sum_{j=1}^r m_j = M\Big\}.$$

Proof: It is an immediate consequence of the definitions that $\forall m > 0$, $\forall i \in I_0$, $E_{(m\delta,i)}$ belongs to the $\mathbb{Q}[q,q^{-1}]$-linear span of

$$\Big\{ \prod_{m_1,\dots,m_r>0} \tilde{E}_{(m_j\delta,i)} \Big| \sum_{j=1}^{r} m_j = m \Big\}$$

and $\tilde{E}_{(m\delta,i)}$ belongs to the $\mathbb{Q}[q,q^{-1}]$-linear span of

$$\Big\{ \prod_{m_1,\dots,m_r>0} E_{(m_j\delta,i)} \Big| \sum_{j=1}^{r} m_j = m \Big\}.$$

The thesis then follows. □

Definition 5.
Let $\prec$ be a linear ordering of $\tilde{R}_+$ and $\forall \alpha \in \tilde{R}_+$ suppose given $x_\alpha \in \mathcal{U}_q$, with $x_\alpha \in \mathcal{U}^+_{q,\mathrm{im}}$ $\forall \alpha \in R^{\mathrm{im}}_+ \times I_0$ (or $x_\alpha \in \mathcal{U}^-_{q,\mathrm{im}}$ $\forall \alpha \in R^{\mathrm{im}}_+ \times I_0$, where $\mathcal{U}^-_{q,\mathrm{im}} \doteqdot \Omega(\mathcal{U}^+_{q,\mathrm{im}})$); then $\forall \underline{\gamma} \doteqdot (\gamma_1,\dots,\gamma_r) \in \cup_{n\leq 0}\tilde{R}^n_+$ I define $x(\underline{\gamma},\prec) \doteqdot x_{\gamma_{\sigma(1)}} \cdot \ldots \cdot x_{\gamma_{\sigma(r)}}$ where $\sigma \in \mathfrak{S}_r$ is such that $\gamma_{\sigma(1)} \preceq \dots \preceq \gamma_{\sigma(r)}$.

Definition 6.
A linear ordering $\prec$ on $\tilde{R}_+$ is said to be a good ordering if

$$\beta_r \preceq \beta_s \Leftrightarrow 0 < r < s \text{ or } r > 0 \geq s \text{ or } r < s \leq 0$$

$$\beta_r \prec \gamma \quad \forall r > 0\ \forall \gamma \in R^{\mathrm{im}}_+ \times I_0$$

$$\gamma \prec \beta_r \quad \forall r \leq 0\ \forall \gamma \in R^{\mathrm{im}}_+ \times I_0.$$

Remark 7.
If $\prec$ and $\prec'$ are good orderings of $\tilde{R}_+$ and $\{x_\alpha | \alpha \in \tilde{R}_+\}$ is as in definition 1.C2.5, then $\forall \underline{\gamma} \in \cup_{n\geq 0}\tilde{R}^n_+$ $x(\underline{\gamma},\prec) = x(\underline{\gamma},\prec')$ and $x(\underline{\gamma},\succ) = x(\underline{\gamma},\succ')$.
Then I give the following definition:

Definition 8.
$\forall \underline{\gamma} \in \cup_{n\geq 0}\tilde{R}^n_+$ I put

$$x(\underline{\gamma}) \doteqdot x(\underline{\gamma},\prec), x(-\underline{\gamma}) \doteqdot x(\underline{\gamma},\succ),$$

where $\prec$ is any good ordering of $\tilde{R}_+$.
I can now state the following theorem

Theorem 9.
The sets

$$\cup_{\eta\in Q_+}\{E(\underline{\gamma}) | \underline{\gamma} \in \mathrm{Par}(\eta)\} = \{E(\underline{\gamma}) | \underline{\gamma} \in \cup_{n\geq 0}(\tilde{R}_+)^n\},$$

$$\cup_{\eta\in Q_+}\{\tilde E(\underline\gamma)|\underline\gamma\in\mathrm{Par}(\eta)\}=\{\tilde E(\underline\gamma)|\underline\gamma\in\cup_{n\geq 0}(\tilde R_+)^n\},$$

$$\cup_{\eta\in Q_+}\{E(-\underline\gamma)|\underline\gamma\in\mathrm{Par}(\eta)\}=\{E(-\underline\gamma)|\underline\gamma\in\cup_{n\geq 0}(\tilde R_+)^n\},$$

$$\cup_{\eta\in Q_+}\{\tilde E(-\underline\gamma)|\underline\gamma\in\mathrm{Par}(\eta)\}=\{\tilde E(-\underline\gamma)|\underline\gamma\in\cup_{n\geq 0}(\tilde R_+)^n\}$$

are bases of $\mathcal{U}_q^+$ and $\cup_{\eta\in Q_+}\mathrm{Par}(\eta)$ is a set of indices for these bases.
More precisely, $\forall\eta\in Q$ the sets $\{E(\underline\gamma)|\underline\gamma\in\mathrm{Par}(\eta)\}$, $\{\tilde E(\underline\gamma)|\underline\gamma\in\mathrm{Par}(\eta)\}$, $\{E(-\underline\gamma)|\underline\gamma\in\mathrm{Par}(\eta)\}$ and $\{\tilde E(-\underline\gamma)|\underline\gamma\in\mathrm{Par}(\eta)\}$ span the same $\mathbb{Q}[q,q^{-1}]$-submodule of $\mathcal{U}_q^+$, which is indeed a $\mathbb{Q}[q,q^{-1}]$-subalgebra of $\mathcal{U}_q^+$.

Proof: The only fact which hasn't yet been proved is that the $\mathbb{Q}[q,q^{-1}]$-linear span of $\{E(-\underline\gamma)|\underline\gamma\in\mathrm{Par}(\eta)\}$ is the same as the $\mathbb{Q}[q,q^{-1}]$-linear span of $\{E(\underline\gamma)|\underline\gamma\in\mathrm{Par}(\eta)\}$, and this fact follows from Levendorskii-Soibelman formulas (see Beck [2]). □

Before studying more closely some properties of the root vectors, I illustrate the case of $\widehat{sl(2)}$:

Example 10.
Let $\mathfrak{g}=\widehat{sl(2)}$. Then the following facts hold:
1) $\rho\check{}=\omega\check{_1}=s_0\tau$ and $2\rho\check{}=s_0s_1$, where τ is the only non trivial Dynkin diagram automorphism: indeed

$$\omega\check{_1}(\alpha_1)=\alpha_1-\delta=-\alpha_0=s_0(\alpha_0)=s_0\tau(\alpha_1);$$

on the other hand $2\rho\check{}=s_0\tau s_0\tau=s_0s_{\tau(0)}=s_0s_1$.
2) $\forall m\geq 0$ $E_{m\delta+\alpha_0}=(T_0T_\tau)^m(E_0)$ and $E_{m\delta+\alpha_1}=(T_0T_\tau)^{-m}(E_1)$: indeed

$$(s_0\tau)^m(\alpha_0)=(\omega\check{_1})^m(\delta-\alpha_1)=\delta-\alpha_1+m\delta=m\delta+\alpha_0,$$

so that $(T_0T_\tau)^m(E_0)=E_{m\delta+\alpha_0}$;
analoglously

$$(s_1\tau)^m(\alpha_1)=(s_0\tau)^{-m}(\alpha_1)=(\omega\check{_1})^{-m}(\alpha_1)=\alpha_1+m\delta,$$

from which $(T_0T_\tau)^{-m}(E_1)=E_{m\delta+\alpha_1}$.
3) $\forall m>0$ $\tilde E_{m\delta}=q^{-2}E_1E_{m\delta-\alpha_1}-E_{m\delta-\alpha_1}E_1=$

$$=q^{-2}E_{r\delta+\alpha_1}E_{(m-r)\delta-\alpha_1}-E_{(m-r)\delta-\alpha_1}E_{r\delta+\alpha_1}\quad\forall r=0,...,m-1$$

(in particular $T_{\omega\check{_1}}(\tilde E_{m\delta})=\tilde E_{m\delta}$ $\forall m>0$): see the definition of $\tilde E_{(m\delta,i)}$ (in this case $I_0=\{1\}$) and Damiani [5].
4) The following commutation rules hold among the root vectors:

$$E_{r\delta+\alpha_1}E_{s\delta+\alpha_0}=q^2E_{s\delta+\alpha_0}E_{r\delta+\alpha_1}+q^2\tilde E_{(r+s+1)\delta}\quad\forall r,s\geq 0,$$

$$[\tilde{E}_{m\delta}, E_{r\delta+\alpha_0}] = -[2]_q\Big(\sum_{s=1}^{m-1} q^{-2(s-1)}(q-q^{-1})E_{(r+s)\delta+\alpha_0}\tilde{E}_{(m-s)\delta}+$$

$$-q^{-2(m-1)}E_{(r+m)\delta+\alpha_0}\Big) \quad \forall r\geq 0,\ \forall m>0,$$

$$[E_{r\delta+\alpha_1}, \tilde{E}_{m\delta}] = -[2]_q\Big(\sum_{s=1}^{m-1} q^{-2(s-1)}(q-q^{-1})\tilde{E}_{(m-s)\delta}E_{(r+s)\delta+\alpha_1}+$$

$$-q^{-2(m-1)}E_{(r+m)\delta+\alpha_1}\Big) \quad \forall r\geq 0,\ \forall m>0,$$

$$[\tilde{E}_{r\delta}, \tilde{E}_{s\delta}] = 0 \quad \forall r,s>0,$$

$$E_{r\delta+\alpha_0}E_{s\delta+\alpha_0} = q^{-2}E_{s\delta+\alpha_0}E_{r\delta+\alpha_0}+$$

$$+\sum_{h=1}^{[\frac{r-s-1}{2}]} q^{-2(h-1)}(q^{-4}-1)E_{(s+h)\delta+\alpha_0}E_{(r-h)\delta+\alpha_0}+$$

$$+\delta_{[\frac{r-s}{2}],\frac{r-s}{2}}q^{-(r-s-2)}(q^{-2}-1)E^2_{\frac{r+s}{2}\delta+\alpha_0} \quad \forall r>s\geq 0,$$

$$E_{s\delta+\alpha_1}E_{r\delta+\alpha_1} = q^{-2}E_{r\delta+\alpha_1}E_{s\delta+\alpha_1}+$$

$$+\sum_{h=1}^{[\frac{r-s-1}{2}]} q^{-2(h-1)}(q^{-4}-1)E_{(r-h)\delta+\alpha_1}E_{(s+h)\delta+\alpha_1}+$$

$$+\delta_{r-s,2[\frac{r-s}{2}]}q^{-(r-s-2)}(q^{-2}-1)E^2_{\frac{r+s}{2}\delta+\alpha_1} \quad \forall r>s\geq 0:$$

see Damiani [5].

Lemma 11.
Let β_r and $\beta_s \in R_+^{\mathrm{re}}$ be such that $\beta_r = \rho^\check{}(\beta_s)$; then either $r,s>0$ or $r,s\leq 0$; moreover $E_{\beta_r} = T_{\rho^\check{}}(E_{\beta_s})$.

Proof: That $r,s>0$ or $r,s\leq 0$ follows from the fact that $\beta_r-\beta_s\in\mathbb{Z}\delta$ and from corollary 1.C1.5.
Suppose $s>0$; then if $\beta_s = w(\alpha_i)$ we have that

$$\beta_r = \rho^\check{}w(\alpha_i) = \rho^\check{}w\tau_n^{-1}(\alpha_{\tau_n(i)})$$

and, by remark 1.C2.1

$$E_{\beta_r} = T_{\rho^\check{}w\tau_n^{-1}}(E_{i+\frac{N}{2}}) = T_{\rho^\check{}}T_wT_{\tau_n^{-1}}(E_{\tau_n(i)}) = T_{\rho^\check{}}T_w(E_i) = T_{\rho^\check{}}(E_{\beta_s}).$$

On the other hand if $r\leq 0$ and $\beta_r = w(\alpha_i)$ it follows that $\beta_s = (\rho^\check{})^{-1}w(\alpha_i) =$

$= (\rho^\vee)^{-1} w\tau_n(\alpha_{i-\frac{N}{2}})$ so that

$$E_{\beta_s} = T^{-1}_{\tau_n^{-1} w^{-1} \rho^\vee}(E_{i-\frac{N}{2}}) =$$

$$= T_{\rho^\vee}^{-1} T_{w^{-1}}^{-1} T_{\tau_n}(E_{\tau_n^{-1}(i)}) = T_{\rho^\vee}^{-1} T_{w^{-1}}^{-1}(E_i) = T_{\rho^\vee}^{-1}(E_{\beta_r}).$$

□

Proposition 12.
$\forall i \in I_0 \ \forall m \in \mathbb{N}$

$$E_{m\delta+\alpha_i} = T_{\omega_i^\vee}^{-m}(E_i)$$

$$E_{(m+1)\delta-\alpha_i} = -T_{\omega_i^\vee}^{m+1}(K_i^{-1}F_i) = T_{\omega_i^\vee}^{m+1} T_i^{-1}(E_i).$$

Proof: Since $\forall m \in \mathbb{N}$

$$(m+1)\delta + \alpha_i = (\rho^\vee)^{-1}(m\delta + \alpha_i), \quad (m+2)\delta - \alpha_i = (\rho^\vee)^{-1}((m+1)\delta - \alpha_i),$$

one has that $\forall m \in \mathbb{N}$

$$E_{m\delta+\alpha_i} = T_{\rho^\vee}(E_{(m+1)\delta+\alpha_i}), \quad E_{(m+2)\delta-\alpha_i} = T_{\rho^\vee}(E_{(m+1)\delta-\alpha_i}).$$

Now it is obvious that $E_{\alpha_i} = E_i$ (see remark 1.B2.8). On the other hand

$$T_{\omega_i^\vee} T_i^{-1}(E_i) = -T_{\omega_i^\vee}(K_i^{-1}F_i) = -\prod_{j \leq i} T_{\omega_j^\vee}(K_i^{-1}F_i) =$$

$$= -T_{\omega_1^\vee} \cdot \ldots \cdot T_{\omega_i^\vee}(K_i^{-1}F_i) = T_{\omega_1^\vee} \cdot \ldots \cdot T_{\omega_i^\vee} T_i^{-1}(E_i) = E_\beta$$

for some $\beta \in R_+^{\mathrm{re}}$, because in the chosen reduced expression of $2\rho^\vee \ \omega_1^\vee \cdot \ldots \cdot \omega_i^\vee =$ $= s_{i_1} \cdot \ldots \cdot s_{i_r}$ for some $r \leq N$, with $i_r = i$; but $\beta = \omega_1^\vee \cdot \ldots \cdot \omega_i^\vee s_i(\alpha_i) = \delta - \alpha_i$, hence $E_{\delta-\alpha_i} = T_{\omega_i^\vee} T_i^{-1}(E_i)$.
Then the thesis follows from the fact that $\forall i \in I_0$, $\forall m > 0$, $T_{\rho^\vee}^m(E_i) = T_{\omega_i^\vee}^m(E_i)$, $T_{\rho^\vee}^m(F_i) = T_{\omega_i^\vee}^m(F_i)$ and $T_{\rho^\vee}^m(K_i) = T_{\omega_i^\vee}^m(K_i)$: indeed $T_{\rho^\vee} = \prod_{i \in I_0} T_{\omega_i^\vee}$ and $\forall j \neq i \ T_{\omega_j^\vee}(E_i) = E_i$, $T_{\omega_j^\vee}(F_i) = F_i$, $T_{\omega_j^\vee}(K_i) = K_i$. □

Definition 13.
$\forall i \in I_0$ define $\mathcal{U}_q(i)$ as the Ω-stable $\mathbb{C}(q_i)$-subalgebra of $\mathcal{U}_q$ generated by

$$\{E_{m\delta+\alpha_i}, E_{(m+1)\delta-\alpha_i}, K_{r\delta\pm\alpha_i} | m \in \mathbb{N}, r \in \mathbb{Z}\}.$$

Of course $E_{(m\delta,i)}$, $\tilde{E}_{(m\delta,i)} \in \mathcal{U}_q(i)$ $\forall m > 0$, thanks to proposition 1.C2.12 and to the definitions of $E_{(m\delta,i)}$, $\tilde{E}_{(m\delta,i)}$.

Theorem 14.
$\mathcal{U}_q(i) \cong \mathcal{U}_{q_i}(\widehat{sl(2)})$ where the isomorphism is given by

$$\Phi_i : \mathcal{U}_{q_i}(\widehat{sl(2)}) \to \mathcal{U}_q(i)$$

$$E_0 \mapsto E_{\delta-\alpha_i}$$

$$E_1 \mapsto E_i$$

$$F_0 \mapsto F_{\delta-\alpha_i}$$

$$F_1 \mapsto F_i$$

$$K_0 \mapsto K_{\delta-\alpha_i}$$

$$K_1 \mapsto K_i$$

(see Damiani [5] for the notations) and the following diagrams are commutative:

$$\begin{array}{ccc} \mathcal{U}_{q_i}(\widehat{sl(2)}) & \longrightarrow & \mathcal{U}_q(i) \\ {\scriptstyle T_1}\big\downarrow & & \big\downarrow{\scriptstyle T_i} \\ \mathcal{U}_{q_i}(\widehat{sl(2)}) & \longrightarrow & \mathcal{U}_q(i) \end{array} \qquad \begin{array}{ccc} \mathcal{U}_{q_i}(\widehat{sl(2)}) & \longrightarrow & \mathcal{U}_q(i) \\ {\scriptstyle \Omega}\big\downarrow & & \big\downarrow{\scriptstyle \Omega} \\ \mathcal{U}_{q_i}(\widehat{sl(2)}) & \longrightarrow & \mathcal{U}_q(i) \end{array}$$

$$\begin{array}{ccc} \mathcal{U}_{q_i}(\widehat{sl(2)}) & \longrightarrow & \mathcal{U}_q(i) \\ {\scriptstyle T_\tau}\big\downarrow & & \big\downarrow{\scriptstyle T_iT_{\omega_i^\vee}^{-1} = T_{\omega_i^\vee}T_i^{-1}} \\ \mathcal{U}_{q_i}(\widehat{sl(2)}) & \longrightarrow & \mathcal{U}_q(i) \end{array} \qquad \begin{array}{ccc} \mathcal{U}_{q_i}(\widehat{sl(2)}) & \longrightarrow & \mathcal{U}_q(i) \\ {\scriptstyle T_0}\big\downarrow & & \big\downarrow{\scriptstyle T_{\omega_i^\vee}^2 T_i^{-1}} \\ \mathcal{U}_{q_i}(\widehat{sl(2)}) & \longrightarrow & \mathcal{U}_q(i) \end{array}$$

where τ is the non trivial Dynkin diagram automorphism of $\widehat{sl(2)}$.

Proof: See Beck [1]. □

Corollary 15.

$\forall i \in I_0$ there is a unique (injective) homomorphism of groups

$$\varphi_i : Q(\widehat{sl(2)}) \to Q$$

such that:

1) $\Phi_i(\mathcal{U}_{q_i,\alpha}(\widehat{sl(2)})) \subseteq \mathcal{U}_{q,\varphi_i(\alpha)}\ \forall \alpha \in Q(\widehat{sl(2)})$;

2) $\varphi_i(R(\widehat{sl(2)})) \subseteq R$;

φ_i is given by $\varphi_i(\alpha_0) = \delta - \alpha_i$, $\varphi_i(\alpha_1) = \alpha_i$; moreover $\forall \alpha \in R_+(\widehat{sl(2)})$

$$\Phi_i(E_\alpha) = \begin{cases} E_{\varphi_i(\alpha)} & \text{if } \alpha \in R_+^{\mathrm{re}}(\widehat{sl(2)}) \\ E_{(\varphi_i(\alpha),i)} & \text{if } \alpha \in R_+^{\mathrm{im}}(\widehat{sl(2)}). \end{cases}$$

Proof: $\varphi_i(\pm\alpha_0) = \pm(\delta - \alpha_i)$ and $\varphi_i(\pm\alpha_1) = \pm\alpha_i$ are necessary conditions for a map $\varphi_i : Q(\widehat{sl(2)}) \to Q$ to be such that $\Phi_i(\mathcal{U}_{q_i,\alpha}(\widehat{sl(2)})) \subseteq \mathcal{U}_{q,\varphi_i(\alpha)}$. Moreover, since $\forall\mathfrak{g}$ Kac-Moody algebra and $\forall\alpha\in Q$

$$\mathcal{U}^+_{q,\alpha}(\mathfrak{g}) = \bigoplus_{\substack{(i_1,\dots,i_r):\\ \sum\alpha_{i_j}=\alpha}} \mathcal{U}^+_{q,\alpha_{i_1}}(\mathfrak{g})\cdot\ldots\cdot\mathcal{U}^+_{q,\alpha_{i_r}}(\mathfrak{g}),$$

$$\mathcal{U}^-_{q,\alpha}(\mathfrak{g}) = \bigoplus_{\substack{(i_1,\dots,i_r):\\ \sum\alpha_{ij}=-\alpha}} \mathcal{U}^-_{q,-\alpha_{i_1}}(\mathfrak{g})\cdot\ldots\cdot\mathcal{U}^-_{q,-\alpha_{i_r}}(\mathfrak{g}),$$

and

$$\mathcal{U}_{q,\alpha}(\mathfrak{g}) = \bigoplus_{\beta\in Q}\mathcal{U}^-_{q,\beta}(\mathfrak{g})\mathcal{U}^0_q(\mathfrak{g})\mathcal{U}^+_{q,\alpha-\beta}(\mathfrak{g}),$$

it follows that $\varphi_i : Q(\widehat{sl(2)}) \to Q$ such that

$$\Phi_i(\mathcal{U}_{q_i,\alpha}(\widehat{sl(2)})) \subseteq \mathcal{U}_{q,\varphi_i(\alpha)}$$

exists and it is the only group homomorphism satisfying the necessary conditions. φ_i is obviously injective and $\varphi_i(\alpha_0 + \alpha_1) = \delta$, from which

$$\varphi_i(R(\widehat{sl(2)})) \subseteq R.$$

Finally, since $\Phi_i T_0 T_\tau = {}^{2}_{\omega_i^\vee}T_i^{-1}\Phi_i T_\tau = T^2_{\omega_i^\vee}T_i^{-1}T_i T^{-1}_{\omega_i^\vee}\Phi_i = T_{\omega_i^\vee}\Phi_i$, $\forall m\in\mathbb{N}$,

$$\Phi_i(E_{m\delta+\alpha_0}) = \Phi_i(T_0T_\tau)^m(E_0) = T^m_{\omega_i^\vee}\Phi_i(E_0) =$$

$$= T^m_{\omega_i^\vee}(E_{\delta-\alpha_i}) = E_{(m+1)\delta-\alpha_i} = E_{\varphi_i(m\delta+\alpha_0)},$$

$$\Phi_i(E_{m\delta+\alpha_1}) = \Phi_i(T_0T_\tau)^{-m}(E_1) = T^{-m}_{\omega_i^\vee}\Phi_i(E_1) =$$

$$= T^{-m}_{\omega_i^\vee}(E_i) = E_{m\delta+\alpha_i} = E_{\varphi_i(m\delta+\alpha_1)},$$

$$\Phi_i(E_{(m+1)\delta}) = \Phi_i(-E_{m\delta+\alpha_0}E_1 + q^{-2}E_1E_{m\delta+\alpha_0}) =$$

$$= -T^{m+1}_{\omega_i^\vee}T_i^{-1}(E_i)E_i + q_i^{-2}E_iT^{m+1}_{\omega_i^\vee}T_i^{-1}(E_i) =$$

$$= T^{m+1}_{\omega_i^\vee}(K_i^{-1}F_i)E_i - q_i^{-2}E_iT^{m+1}_{\omega_i^\vee}(K_i^{-1}F_i) =$$

$$= E_{((m+1)\delta,i)} = E_{(\varphi_i((m+1)\delta),i)}.$$

□

Part D. Integer forms, Verma modules and specialization.
I want now to consider q as a complex parameter, that is $\forall\varepsilon$ in a suitable subset of $\mathbb{C}$ I want to find a $\mathbb{C}$-algebra $\mathcal{U}_\varepsilon(\mathfrak{g})$ obtained by "specializing" q at ε.

To this aim I introduce the integer forms of $\mathcal{U}_q(\mathfrak{g})$ and the specialization at a nonzero complex parameter ε.

§D1. Integer form and specialization of $\mathcal{U}_q(\hat{\mathfrak{g}})$.

Definition 1.
$\mathcal{A} \doteq \mathbb{C}[q, q^{-1}, (q^r - q^{-r})^{-1} | 1 \leq r \leq \max\{d_i | i \in I_0\}]$.

Definition 2.
If $\tilde{\mathcal{A}}$ is a ring such that $\mathbb{C}[q, q^{-1}] \subseteq \tilde{\mathcal{A}} \subseteq \mathbb{C}(q)$, denote by $\mathcal{U}_{\tilde{\mathcal{A}}}$ the $\tilde{\mathcal{A}}$-subalgebra of $\mathcal{U}_q$ generated by $\{E_i, F_i, K_i^{\pm 1} | i \in I\}$.
Define also the following $\tilde{\mathcal{A}}$-subalgebras of $\mathcal{U}_{\tilde{\mathcal{A}}}$:
$\mathcal{U}_{\tilde{\mathcal{A}}}^+$ and $\mathcal{U}_{\tilde{\mathcal{A}}}^-$ are the subalgebras generated by $\{E_i | i \in I\}$ and $\{F_i | i \in I\}$ respectively; $\mathcal{U}_{\tilde{\mathcal{A}}}^0$ is the subalgebra generated by $\left\{K_i^{\pm 1}, \frac{K_i - K_i^{-1}}{q_i - q_i^{-1}} \Big| i \in I\right\}$.
Moreover $\forall \alpha \in Q$ one can define the following $\tilde{\mathcal{A}}$-submodules of $\mathcal{U}_{\tilde{\mathcal{A}}}$:

$$\mathcal{U}_{\tilde{\mathcal{A}},\alpha} \doteq \mathcal{U}_{\tilde{\mathcal{A}}} \cap \mathcal{U}_{q,\alpha}, \quad \mathcal{U}_{\tilde{\mathcal{A}},\alpha}^+ \doteq \mathcal{U}_{\tilde{\mathcal{A}},\alpha} \cap \mathcal{U}_{\tilde{\mathcal{A}}}^+, \quad \mathcal{U}_{\tilde{\mathcal{A}},\alpha}^- \doteq \mathcal{U}_{\tilde{\mathcal{A}},\alpha} \cap \mathcal{U}_{\tilde{\mathcal{A}}}^-.$$

Remark 3.
If $\tilde{\mathcal{A}}$ is such that $(q_i - q_i^{-1})^{-1} \in \tilde{\mathcal{A}}\ \forall i \in I$, then $\mathcal{U}_{\tilde{\mathcal{A}}}^0 \cong \tilde{\mathcal{A}}[K_1^{\pm 1}, ..., K_n^{\pm 1}]$. □

Proposition 4.

$$\mathcal{U}_{\tilde{\mathcal{A}}} = \bigoplus_{\alpha \in Q} \mathcal{U}_{\tilde{\mathcal{A}},\alpha}, \quad \mathcal{U}_{\tilde{\mathcal{A}}}^+ = \bigoplus_{\alpha \in Q_+} \mathcal{U}_{\tilde{\mathcal{A}},\alpha}^+, \quad \mathcal{U}_{\tilde{\mathcal{A}}}^- = \bigoplus_{\alpha \in Q_+} \mathcal{U}_{\tilde{\mathcal{A}},-\alpha}^-.$$

□

Proposition 5.
$\mathcal{U}_{\tilde{\mathcal{A}}} \cong \mathcal{U}_{\tilde{\mathcal{A}}}^- \otimes \mathcal{U}_{\tilde{\mathcal{A}}}^0 \otimes \mathcal{U}_{\tilde{\mathcal{A}}}^+$ (triangular decomposition for the integer form).
In particular $\mathcal{U}_{\tilde{\mathcal{A}}}^+ = \mathcal{U}_{\tilde{\mathcal{A}}} \cap \mathcal{U}_q^+$, $\mathcal{U}_{\tilde{\mathcal{A}}}^- = \mathcal{U}_{\tilde{\mathcal{A}}} \cap \mathcal{U}_q^-$ and $\mathcal{U}_{\tilde{\mathcal{A}}}^0 = \mathcal{U}_{\tilde{\mathcal{A}}} \cap \mathcal{U}_q^0$.
Moreover $\mathcal{U}_{\tilde{\mathcal{A}}} \cong \mathcal{U}_{\tilde{\mathcal{A}}}^- \otimes \tilde{\mathcal{A}}[K_i^{\pm 1} | i \in I] \otimes \mathcal{U}_{\tilde{\mathcal{A}}}^+ \Leftrightarrow (q_i - q_i^{-1})^{-1} \in \tilde{\mathcal{A}}\ \forall i \in I$. □

Proposition 6.
$\mathcal{U}_{\tilde{\mathcal{A}}}$ is Ω-stable if and only if $\tilde{\mathcal{A}}$ is Ω-stable.
Moreover $\mathcal{U}_{\tilde{\mathcal{A}}}$ is T_i-stable if and only if $([\max_{j \in I}(-a_{i,j})]_{q_i}!)^{-1} \in \tilde{\mathcal{A}}$, and this happens if and only if it is T_i^{-1}-stable (in particular if $\mathcal{U}_{\tilde{\mathcal{A}}}$ is T_i-stable, then $T_i \in \mathrm{Aut}_{\tilde{\mathcal{A}}}(\mathcal{U}_{\tilde{\mathcal{A}}})$).
Finally $\mathcal{U}_{\tilde{\mathcal{A}}}$ is T_τ-stable $\forall \tau$ automorphism of the Dynkin diagram. □

Corollary 7.
Ω, $T_i^{\pm 1}$ $(i \in I)$ and T_τ $(\tau \in \mathcal{T})$ act on $\mathcal{U}_{\mathcal{A}}$; moreover $\mathcal{U}_{\mathcal{A}} \cong \mathcal{U}_{\mathcal{A}}^- \otimes \mathcal{U}_{\mathcal{A}}^0 \otimes \mathcal{U}_{\mathcal{A}}^+$ with $\mathcal{U}_{\mathcal{A}}^0 \cong \mathcal{A}[K_1^{\pm 1}, ..., K_n^{\pm 1}]$.

Definition 8.
If $\tilde{\mathcal{A}}$ is a subring of $\mathbb{C}(q)$ containing $\mathbb{C}[q,q^{-1}]$, $\tilde{\mathcal{U}}_{\tilde{\mathcal{A}}}$ denotes the least $\tilde{\mathcal{A}}$-subalgebra of $\mathcal{U}_q$ containing $\mathcal{U}_{\tilde{\mathcal{A}}}$ and stable under the action of the braid group.

Remark 9.
$\tilde{\mathcal{U}}_{\tilde{\mathcal{A}}}$ is stable under the action of the extended braid group.

Proof: It follows from the fact that $T_\tau(\mathcal{U}_{\tilde{\mathcal{A}}}) \subseteq \mathcal{U}_{\tilde{\mathcal{A}}}$. □

Remark 10.
If $\tilde{\mathcal{A}} \supseteq \mathcal{A}$ we have that $\tilde{\mathcal{U}}_{\tilde{\mathcal{A}}} = \mathcal{U}_{\tilde{\mathcal{A}}}$ (because $\mathcal{U}_{\tilde{\mathcal{A}}}$ is $\tilde{\mathcal{B}}$-stable).

Proposition 11.
If A is of untwisted affine type, $\tilde{\mathcal{U}}_{\mathbb{C}[q,q^{-1}]}$ is the $\mathbb{C}[q,q^{-1}]$-subalgebra of $\mathcal{U}_q$ generated by $\{E_\alpha, F_\alpha, K_i^{\pm 1}\}$ and the monomials $F(\underline{\gamma})K_\lambda E(\tilde{\underline{\gamma}})$ form a $\mathbb{C}[q,q^{-1}]$-basis of $\tilde{\mathcal{U}}_{\mathbb{C}[q,q^{-1}]}$. □

Definition 12.
$\forall \varepsilon \in \mathbb{C}^*$ I call specialization of $\mathcal{U}_q$ at ε the $\mathbb{C}$-algebra $\mathcal{U}_\varepsilon = \mathcal{U}_\varepsilon(\mathfrak{g})$ defined by

$$\mathcal{U}_\varepsilon \doteqdot \tilde{\mathcal{U}}_{\mathbb{C}[q,q^{-1}]}/(q-\varepsilon).$$

Remark 13.
If $\varepsilon^r \neq \varepsilon^{-r}$ $\forall r = 1, ..., \max\{d_i\}$ then $\mathcal{U}_\varepsilon \cong \mathcal{U}_{\mathcal{A}}/(q-\varepsilon)$.
In particular $\mathcal{U}_\varepsilon$ is $\tilde{\mathcal{B}}$-stable. □

Definition 14.
$\mathbb{C}$-subalgebras and $\mathbb{C}$-vector subspaces of $\mathcal{U}_\varepsilon$ $\mathcal{U}_\varepsilon^+$, $\mathcal{U}_\varepsilon^-$, $\mathcal{U}_\varepsilon^0$, $\mathcal{U}_{\varepsilon,\alpha}$, $\mathcal{U}_{\varepsilon,\alpha}^+$, $\mathcal{U}_{\varepsilon,\alpha}^-$,... are defined as the images of the corresponding $\mathbb{C}[q,q^{-1}]$-subalgebras and $\mathbb{C}[q,q^{-1}]$-submodules of $\tilde{\mathcal{U}}_{\mathbb{C}[q,q^{-1}]}$.

Remark 15.
If $\varepsilon^{d_i} \neq \varepsilon^{-d_i}$ $\forall i \in I$ there is for $\mathcal{U}_\varepsilon$ the triangular decomposition

$$\mathcal{U}_\varepsilon \cong \mathcal{U}_\varepsilon^- \otimes_{\mathbb{C}} \mathcal{U}_\varepsilon^0 \otimes_{\mathbb{C}} \mathcal{U}_\varepsilon^+$$

where $\mathcal{U}_\varepsilon^0 \cong \mathbb{C}[K_i^{\pm 1} | i \in I]$.

Proof: The triangular decomposition of $\mathcal{U}_\varepsilon$ is that induced by the triangular decomposition of $\tilde{\mathcal{U}}_{\mathbb{C}[q,q^{-1}]}$. □

Remark 16.
The (anti)automorphisms of $\mathcal{U}_q$ Ω, T_i and T_τ induce analogous (anti)automorphisms on $\mathcal{U}_\varepsilon$ under the following conditions.
T_τ is always defined as a $\mathbb{C}$-automorphism of $\mathcal{U}_\varepsilon$.

Ω is defined if $\Omega(q-\varepsilon) = q^{-1} - \bar{\varepsilon} \in (q-\varepsilon)$ that is if and only if $q - \bar{\varepsilon}^{-1} \in (q-\varepsilon)$; this happens exactly when $\bar{\varepsilon} = \varepsilon^{-1}$, that is when $|\varepsilon| = 1$. In particular Ω is defined as $\mathbb{C}$-antilinear antiinvolution of $\mathcal{U}_\varepsilon$ when ε is a root of unity.
Finally T_i is defined as automorphism of $\mathcal{U}_\varepsilon$ if the specialization at ε of $[\max_{j\in I}(-a_{i,j})]_{q_i}!$ is different from zero, that is if

$$\frac{\varepsilon^{rd_i} - \varepsilon^{-rd_i}}{\varepsilon^{d_i} - \varepsilon^{-d_i}} \neq 0 \ \ \forall r = 1, ..., \max_j(-a_{i,j});$$

this means that $\forall i \in I \ \forall r \in \{1, ..., \max_j(-a_{i,j})\} \ \varepsilon^{2rd_i} = 1 \Rightarrow \varepsilon^{2d_i} = 1$.
Object of this thesis is a particular class of specializations of $\mathcal{U}_q$, namely the specializations at primitive l^{th} roots of 1 under suitable conditions on l. I summarize the preceding considerations in the following proposition:

Proposition 17.
Let $l > 0$ be such that $\forall i \in I \ \forall r \in \{1, ..., \max_j(-a_{i,j})\} \ l \nmid 2rd_i$ and let ε be a primitive l^{th} root of 1. Then the specialization $\mathcal{U}_\varepsilon$ of $\mathcal{U}_q$ at ε is provided with
1) triangular decomposition: $\mathcal{U}_\varepsilon \cong \mathcal{U}_\varepsilon^- \otimes_{\mathbb{C}} \mathcal{U}_\varepsilon^0 \otimes_{\mathbb{C}} \mathcal{U}_\varepsilon^+$ with $\mathcal{U}_\varepsilon^0 \cong \mathbb{C}[K_i^{\pm 1} | i \in I]$;
2) $\mathbb{C}$-antilinear antiinvolution $\Omega : \mathcal{U}_\varepsilon \to \mathcal{U}_\varepsilon$,
3) decomposition $\mathcal{U}_\varepsilon = \oplus_{\alpha \in Q} \mathcal{U}_{\varepsilon,\alpha}$;
4) action of the extended braid group:

$$\forall w \in \tilde{W} \ \ T_w(\mathcal{U}_{\varepsilon,\alpha}) \subseteq \mathcal{U}_{\varepsilon,w(\alpha)} \ \ \forall \alpha \in Q.$$

□

Actually I shall consider slightly more restrictive conditions on l by requiring that l be odd.
Hence the condition $l \nmid 2rd_i \ \forall i \in I \ \forall r \in \{1, ..., \max_j(-a_{i,j})\}$ can be written as $l > \max_{ij\in I}(-a_{i,j})$ (because $d_i \in \{1,2\}$ if $\mathfrak{g} \neq G_2$, and, if $\mathfrak{g} = G_2$, then either $d_i = 1$ and $\max_{j\in I}(-a_{i,j}) = 3$ or $d_i = 3$ and $\max_{j\in I}(-a_{i,j}) = 1$).
To sum up I shall work with an odd number $l > \max_{ij\in I}(-a_{i,j})$ (that is an odd number bigger than 1 if $\mathfrak{g} \neq G_2$, and an odd number bigger than 3 if $\mathfrak{g} = G_2$). Unless otherwise stated l will satisfy these conditions.
I shall now define a basis (indeed a PBW basis) of $\mathcal{U}_\varepsilon$.

Notation 18.
$\forall x \in \mathcal{U}_q$ I denote again by x its projection in $\mathcal{U}_\varepsilon$, if it exists; that is, if $x \in \mathcal{U}_{\mathcal{A}_\varepsilon}$ where $\mathcal{A}_\varepsilon \subset \mathbb{C}(q)$ is the local ring of the rational functions with no poles at ε ($\mathcal{A}_\varepsilon \subset \mathbb{C}(q)$ is the localization of $\mathbb{C}[q, q^{-1}]$ at the maximal ideal $(q-\varepsilon)$).
Equivalently this condition means that there exists an invertible $a \in \mathcal{A}_\varepsilon$ such that $ax \in \mathcal{U}_{\mathbb{C}[q,q^{-1}]}$.

Remark 19.
If ε is a primitive l^{th} root of 1 ($l > \max_{ij\in I}(-a_{i,j})$ odd) then $\forall \alpha \in \tilde{R}_+ \ E_\alpha$, $\tilde{E}_\alpha$, F_α, $\tilde{F}_\alpha \in \mathcal{U}_\varepsilon$.

Moreover $\forall \underline{\beta} \in \cup_{n\in\mathbb{N}}(\tilde{R}_+)^r$ $E(\underline{\beta})$, $\tilde{E}(\underline{\beta})$, $E(-\underline{\beta})$, $\tilde{E}(-\underline{\beta})$, $F(\underline{\beta})$, $\tilde{F}(\underline{\beta})$, $F(-\underline{\beta})$, $\tilde{F}(-\underline{\beta}) \in \mathcal{U}_\varepsilon$. □

Theorem 20.
$\forall \eta \in Q$

$$\{E(\underline{\beta}) | \underline{\beta} \in \mathrm{Par}(\eta)\}, \quad \{\tilde{E}(\underline{\beta}) | \underline{\beta} \in \mathrm{Par}(\eta)\},$$

$$\{E(-\underline{\beta}) | \underline{\beta} \in \mathrm{Par}(\eta)\}, \quad \{\tilde{E}(-\underline{\beta}) | \underline{\beta} \in \mathrm{Par}(\eta)\}$$

are bases of $\mathcal{U}^+_{\varepsilon,\eta}$.
Proof: It follows from proposition 1.D1.11. □

Remark 21.
$\mathcal{U}_{\mathbb{C}[q,q^{-1}]}/(q-1)$ is such that $K_i^2 = 1$ $\forall i \in I$ and

$$\mathcal{U}_{\mathbb{C}[q,q^{-1}]}/(q-1, K_i - 1 | i \in I) \cong \mathcal{U}(\hat{\mathfrak{g}})$$

$$E_i \mapsto e_i$$

$$F_i \mapsto f_i$$

$$\frac{K_i - K_i^{-1}}{q_i - q_i^{-1}} \mapsto h_i.$$

Proof: See De Concini Kac [6]. □

§D2. Verma modules and integer forms.

In this section I'll introduce the notion of Verma module, the Casimir operator and the contravariant form H, which will be the main tool in the study of the center of $\mathcal{U}_\varepsilon$; in particular the calculation of the highest term of its determinant will be the argument of chapter 2.
To this aim I first extend the field of scalars $\mathbb{C}(q)$ to an algebraic closure $\mathbb{k}$ of $\mathbb{C}(q)$: in this way the algebra of polynomials $\mathbb{k}[K_0^{\pm 1}, ..., K_n^{\pm 1}]$ is the $\mathbb{k}$-algebra of polynomial functions on the algebraic variety $(\mathbb{k}^*)^{n+1}$, so that one can apply the results relative to the algebraic varieties on algebraically closed fields.

Definition 1.
Let $\mathbb{k}$ be an algebraic closure of $\mathbb{C}(q)$.
I denote by $\mathcal{U}_{\mathbb{k}}$ the $\mathbb{k}$-algebra $\mathcal{U}_{\mathbb{k}} \doteqdot \mathcal{U}_q \otimes_{\mathbb{C}(q)} \mathbb{k}$.

Remark 2.
I remark that all the results valid for $\mathcal{U}_q$ also hold for $\mathcal{U}_{\mathbb{k}}$, and in particular there are for $\mathcal{U}_{\mathbb{k}}$ the triangular decomposition, the Q-gradation, the action of the extended braid group $\tilde{\mathcal{B}}$ and a $\mathbb{C}$-antiautomorphism Ω such that $\Omega|_{\mathbb{k}}$ verifies the following conditions:

$$\Omega|_{\mathbb{R}} = \mathrm{id}_{\mathbb{R}}, \quad \Omega(i) = -i, \quad \dot{\Omega}(q) = q^{-1}.$$

Note that in particular

$$\mathcal{U}_{\mathbb{k}}^0 = \mathbb{C}(q)[K_0^{\pm 1}, ..., K_n^{\pm 1}] \otimes_{\mathbb{C}(q)} \mathbb{k} = \mathbb{k}[K_0^{\pm 1}, ..., K_n^{\pm 1}].$$

Proof: It is clear that $\Omega|_{\mathbb{C}(q)}$ can be extended to an automorphism of $\mathbb{k}$; then everything follows immediately. □

Definition 3.
$\forall \varphi = (\varphi_0, ..., \varphi_n) \in (\mathbb{k}^*)^{n+1}$ define a structure $\mathbb{k}_\varphi$ of $\mathcal{U}_{\mathbb{k}}^{\geq 0}$-module on $\mathbb{k}$ by setting $E_i.1 \doteq 0$, $K_i.1 \doteq \varphi_i$ $\forall i = 0, ..., n$.

Definition 4.
Let $\varphi \in (\mathbb{k}^*)^{n+1}$; then V^φ denotes the $\mathcal{U}_{\mathbb{k}}$-module induced by $\mathbb{k}_\varphi$.
In the following proposition I describe V^φ in more details.

Proposition 5.
There is an isomorphism of $\mathcal{U}_{\mathbb{k}}^-$-modules between $\mathcal{U}_{\mathbb{k}}^-$ and V^φ; I denote by v_φ the image of 1 in V^φ.
In particular $V^\varphi = \{f.v_\varphi | f \in \mathcal{U}_{\mathbb{k}}^-\}$ is a free $\mathcal{U}_{\mathbb{k}}^-$-module of rank 1, so that

$$\{F(\underline{\gamma}).v_\varphi | \cup_{\eta \in Q_+} \underline{\gamma} \in \mathrm{Par}(\eta)\} \quad \text{and} \quad \{F(-\underline{\gamma}).v_\varphi | \cup_{\eta \in Q_+} \underline{\gamma} \in \mathrm{Par}(\eta)\}$$

are $\mathbb{k}$-bases of V^φ.
Moreover $x.v_\varphi = \tilde{\pi}(x).v_\varphi = \tilde{\pi}(x)(\varphi)v_\varphi$ where $FK_\lambda(\varphi) = K_\lambda(\varphi)F$, so that

$$F_i.(f.v_\varphi) = (F_i f).v_\varphi, \ \forall f, \tilde{f} \in \mathcal{U}_{\mathbb{k}}^-,$$

$$K_i.(f.v_\varphi) = q^{-(\alpha_i|\alpha)}\varphi_i f.v_\varphi, \ \forall i \in I, {}_J \in \mathcal{U}_{\mathbb{k},-\alpha}^-,$$

$$E_i.(f.v_\varphi) = [E_i, f].v_\varphi \ \forall i \in I, f \in \mathcal{U}_{\mathbb{k},-\alpha}^-.$$

Proof: The assertions follow from the fact that

$$V^\varphi = \mathcal{U}_{\mathbb{k}} \otimes_{\mathcal{U}_{\mathbb{k}}^{\geq 0}} \mathbb{k}_\varphi = \mathcal{U}_{\mathbb{k}}^- \otimes_{\mathbb{k}} \mathcal{U}_{\mathbb{k}}^{\geq 0} \otimes_{\mathcal{U}_{\mathbb{k}}^{\geq 0}} \mathbb{k}_\varphi = \mathcal{U}_{\mathbb{k}}^- \otimes_{\mathbb{k}} \mathbb{k}_\varphi \cong \mathcal{U}_{\mathbb{k}}^-.$$

□

Definition 6.
Define the following subspaces of V^φ:
1) V_q^φ is the $\mathbb{C}(q)$-vector subspace of V^φ generated by

$$\{F(\underline{\gamma}).v_\varphi | \cup_{\eta \in Q_+} \underline{\gamma} \in \mathrm{Par}(\eta)\},$$

which is of course a $\mathbb{C}(q)$-basis of V_q^φ; of course also

$$\{F(-\underline{\gamma}).v_\varphi | \cup_{\eta \in Q_+} \underline{\gamma} \in \mathrm{Par}(\eta)\}$$

is a $\mathbb{C}(q)$-basis of V_q^φ;
2) $V_{\mathcal{A}}^\varphi$ is the $\mathcal{A}$-submodule of V^φ generated by $\{F(\underline{\gamma}).v_\varphi | \cup_{\eta\in Q_+} \underline{\gamma}\in \mathrm{Par}(\eta)\}$, which is of course an $\mathcal{A}$-basis of $V_{\mathcal{A}}^\varphi$; of course also

$$\{F(-\underline{\gamma}).v_\varphi | \cup_{\eta\in Q_+} \underline{\gamma}\in \mathrm{Par}(\eta)\}$$

is an $\mathcal{A}$-basis of $V_{\mathcal{A}}^\varphi$;
3) $\tilde{V}^\varphi$ is the $\mathbb{C}[q,q^{-1}]$-submodule of V^φ generated by

$$\{F(\underline{\gamma}).v_\varphi | \cup_{\eta\in Q_+} \underline{\gamma}\in \mathrm{Par}(\eta)\},$$

which is of course a $\mathbb{C}[q,q^{-1}]$-basis of $\tilde{V}^\varphi$; of course also

$$\{F(-\underline{\gamma}).v_\varphi | \cup_{\eta\in Q_+} \underline{\gamma}\in \mathrm{Par}(\eta)\}$$

is a $\mathbb{C}[q,q^{-1}]$-basis of $\tilde{V}^\varphi$.

Remark 7.
$\tilde{V}^\varphi \subseteq V_{\mathcal{A}}^\varphi \subseteq V_q^\varphi \subseteq V^\varphi$. □

Remark 8.
1) V_q^φ is a $\mathcal{U}_q$-module if and only if $\varphi\in(\mathbb{C}(q)^*)^{n+1}$;
2) $V_{\mathcal{A}}^\varphi$ is a $\mathcal{U}_{\mathcal{A}}$-module if and only if $\varphi_i\in\mathcal{A}$ is invertible in $\mathcal{A}$ $\forall i=0,...,n$;
3) $\tilde{V}^\varphi$ is a $\mathcal{U}_{\mathbb{C}[q,q^{-1}]}$-module if and only if $\varphi\in(\mathbb{C}[q,q^{-1}])^{n+1}$ with φ_i invertible in $\mathbb{C}[q,q^{-1}]$ $\forall i=0,...,n$, that is if and only if $\forall i=0,...,n$ φ_i is of the form $a_i q^{r_i}$ with $a_i\in\mathbb{C}^*$, $r_i\in\mathbb{Z}$.
Proof: The claim immediately follows from the action of $\mathcal{U}_{\mathbb{k}}$ on V^φ. □

Definition 9.
Fix $\underline{\delta}\in\{\pm1\}^{n+1}$ and $\underline{r}\in\mathbb{Z}^{n+1}$ and let $\varphi\in(\mathbb{k}^*)^{n+1}$ be defined by $\varphi_i \doteqdot \delta_i q^{r_i}$ $\forall i=0,...,n$. Then the specialization $V_\varepsilon^{\underline{\delta},\underline{r}}$ at $\varepsilon\in\mathbb{C}^*$ of V^φ is $\tilde{V}^\varphi/(q-\varepsilon)\tilde{V}^\varphi$.

Remark 10.
$V_\varepsilon^{\underline{\delta},\underline{r}}$ inherits from the structure of $\mathcal{U}_{\mathbb{k}}$-module on V^φ a structure of $\mathcal{U}_\varepsilon$-module.

Remark 11.
V^φ and its submodules and quotient defined above inherit from $\mathcal{U}_{\mathbb{k}}$ a Q-gradation: $V^\varphi=\oplus_{\alpha\in Q_-}V_{-\alpha}^\varphi$, where $\forall\alpha\in Q$ $V_{-\alpha}^\varphi$ is the image in V^φ of $\mathcal{U}_{\mathbb{k},-\alpha}^-$. Notice that $\forall\alpha\in Q$ $\forall v\in V_{-\alpha}^\varphi$ and $\forall i\in I$, $K_i.v=q^{-(\alpha_i|\alpha)}\varphi_i v$.

Lemma 12.
$\dim_{\mathbb{k}} V_0^\varphi = 1$.
Proof: Indeed $\mathrm{par}(0)=1$. □

Proposition 13.
There exists a unique maximal graded $\mathcal{U}_{\mathbb{k}}$-submodule of V^φ, say N^φ; the quotient $L^\varphi \doteqdot V^\varphi/N^\varphi$ is the unique graded irreducible quotient of V^φ.

Proof: It is enough to define N^φ as the sum of all the graded $\mathcal{U}_{\mathbb{k}}$-submodules of V^φ different from V^φ. This is easily shown to be maximal and Q-graded, and the claim follows. □

Definition 14.
Let $\varphi \in (\mathbb{k}^*)^{n+1}$; an element $v \in V^\varphi$ is said to be primitive if $E_i.v = 0$ $\forall i = 0, ..., n$.

Remark 15.
Remark that a homogeneous primitive vector which does not belong to N^φ is a scalar multiple of v_φ.

§D3. Casimir operator.

In this section I shall introduce the Casimir operator and state some of its properties that I'll need; all the results can be found in Tanisaki [23].

Lemma 1.
Let $v \in V^\varphi$; then there exists $\alpha \in Q$ such that $\sum_{\beta \not\leq \alpha} \mathcal{U}_{\mathbb{k},\beta}.v = 0$.
In particular if $v \in V^\varphi_{-\gamma}$ we have that $\alpha = \gamma$.

Proof: Of course there exists α such that $v \in \sum_{\beta \leq \alpha} V^\varphi_{-\beta}$ and since $\{\beta \leq \alpha\}$ is finite I can suppose that $v \in V^\varphi_{-\alpha}$.
Then if $x \in \mathcal{U}_{\mathbb{k},\beta}$ with $\beta \not\leq \alpha$, $x.v \in V^\varphi_{-\alpha+\beta} = \{0\}$ because $\alpha - \beta \notin Q_+$. □

Corollary 2.
$\forall \alpha \in Q$ let $x_\alpha \in \mathcal{U}_{\mathbb{k}} \mathcal{U}^+_{\mathbb{k},\alpha}$. Then $\sum_{\alpha \in Q} x_\alpha$ defines a $\mathbb{k}$-linear operator on V^φ $\forall \varphi \in (\mathbb{k}^*)^n$.

Proof: It follows immediately from lemma 1.D3.1. □

Definition 3.
$\forall \eta \in Q_+$ let C_η be the canonical element of the pairing $(\cdot,\cdot)|_{\mathcal{U}^+_{\mathbb{k},\eta} \times \mathcal{U}^-_{\mathbb{k},-\eta}}$ (see remark 1.B3.10), which means

$$C_\eta = \sum_{r=1}^{\mathrm{par}(\eta)} x_r \otimes y_r \in \mathcal{U}_{\mathbb{k}} \otimes \mathcal{U}_{\mathbb{k}}$$

where $\{x_r\}$ and $\{y_r\}$ are dual bases with respect to $(\cdot,\ )$, that is $(x_r, y_s) = \delta_{rs}$. Remark that the element C_η is independent of the choice of the dual bases $\{x_r\}$ and $\{y_r\}$.

Remark 4.
$\forall \eta \in Q_+$ $\mathbf{m}(S \otimes \mathrm{id})\sigma(C_\eta) \in \mathcal{U}_{\mathbb{k}} \mathcal{U}^+_{\mathbb{k},\eta}$.

Proof: It immediately follows from the fact that $x_r \in \mathcal{U}^+_{\mathbb{k},\eta}$ $\forall r = 1, ..., \mathrm{par}(\eta)$.

Definition 5.
Define the Casimir operator on the Verma modules to be

$C \doteqdot \sum_{\eta\in Q_+} \mathbf{m}(S\otimes \mathrm{id})\sigma(C_\eta)$.
It is well defined, thanks to corollary 1.D3.2.

Proposition 6.
1) If $v\in V^\varphi$ is a primitive vector, then $C.v=v$;
2) $\forall i=0,...,n$ $CE_i = K_i^{-2}E_iC$ and $CF_i = F_iK_i^2C$; more generally if $\hat\rho\in E^*$ is defined by $\langle\hat\rho,\alpha_i\rangle \doteqdot d_i \quad \forall i=0,...,n$ then

$$Cf_\alpha = q^{2\langle\hat\rho,\alpha\rangle-(\alpha|\alpha)} f_\alpha K_{2\alpha}C \quad \forall f_\alpha \in \mathcal{U}^-_{\Bbbk,-\alpha}.$$

Proof: See Tanisaki [23]. □

§D4. The contravariant form.

On each Verma module V^φ a contravariant form H^φ can be defined; it is the evaluation at φ of a "universal" contravariant form H defined on $\mathcal{U}^-_\Bbbk$ with values in $\mathcal{U}^0_\Bbbk = \Bbbk[K_0^{\pm1},...,K_n^{\pm1}]$. It is in the study of H that the extension of the field of scalars $\mathbb{C}(q)$ to $\Bbbk$ will be clear to be useful.

Proposition 1.
There exists a unique form $H^\varphi : V^\varphi\times V^\varphi$ such that
1) $H^\varphi(v_\varphi,v_\varphi)=1$;
2) $H^\varphi(v+\tilde v,w+\tilde w) = H^\varphi(v,w)+H^\varphi(v,\tilde w)+H^\varphi(\tilde v,w)+H^\varphi(\tilde v,\tilde w)$ $\forall v,\tilde v,w,\tilde w$.
3) $H^\varphi(x.v,w) = H^\varphi(v,\Omega(x).w)$ $\forall v,w\in V^\varphi$, $\forall x\in\mathcal{U}_\Bbbk$.

Proof: H^φ is, if it exists with the stated properties, unique: if $v\in V^\varphi_{-\alpha}$ with $\alpha>0$ then $v=x.v_\varphi$ with $x\in\mathcal{U}^-_{\Bbbk,-\alpha}$, so that

$$H^\varphi(v_\varphi,v) = H^\varphi(v_\varphi,x.v_\varphi) = H^\varphi(\Omega^{-1}(x).v_\varphi,v_\varphi) = H^\varphi(0,v_\varphi)=0$$

This means that $H^\varphi|_{\Bbbk v_\varphi\times\mathcal{U}_\Bbbk}$ is uniquely determined; moreover, let $v,w\in V^\varphi$ and let $x\in\mathcal{U}^-_\Bbbk$ be such that $v=x.v_\varphi$; then necessarily $H^\varphi(v,w) = H^\varphi(x.v_\varphi,w) = H^\varphi(v_\varphi,\Omega(x).w)$ which is, as just proved, uniquely determined.
For the existence of H^φ it is immediate to prove that if H^φ is as above, that is $H^\varphi(v,w)\doteqdot H^\varphi(v_\varphi,(\Omega(x).w)_0)$ where
$v=x.v_\varphi$, $w\in V^\varphi$ and $\Omega(x).w=\sum_{\alpha\in Q_+}(\Omega(x).w)_{-\alpha}$ with $(\Omega(x).w)_{-\alpha}\in V^\varphi_{-\alpha}$, then H^φ is well defined and properties 1), 2) and 3) are obviously satisfied.
□

Definition 2.
The form H^φ is called the contravariant form of V^φ.

Proposition 3.
H^φ has the following properties:
i) $H^\varphi|_{V^\varphi_{-\eta}\times V^\varphi_{-\tilde\eta}}=0$ if $\eta\neq\tilde\eta$;
ii) kerH^φ is N^φ, the (unique) maximal graded $\mathcal{U}_\Bbbk$-submodule of V^φ;

iii) $\ker H^\varphi \cap V^\varphi_{-\eta} \neq \{0\} \Leftrightarrow$ there exists a primitive vector $v \in V^\varphi_{-\alpha}$ where $0 < \alpha \leq \eta$;
iv) $H^\varphi(fv_\varphi, \tilde{f}v_\varphi) = \pi(\Omega(f)\tilde{f})(\varphi)$. ;

Proof: i) follows from the definition of H^φ and implies that $\ker H^\varphi$ (which is a submodule of V^φ) is graded, hence, since $v_\varphi \notin \ker H^\varphi$, $\ker H^\varphi \subseteq N^\varphi$.
Let $v \in N^\varphi$; then v generates a $\mathcal{U}_{\mathbb{k}}$-submodule of V^φ contained in $\sum_{\alpha>0} V^\varphi_{-\alpha}$, so that $H^\varphi(x.v_\varphi, v) = H^\varphi(v_\varphi, \Omega(x).v) = 0 \ \forall x \in \mathcal{U}_{\mathbb{k}}$. Thus $v \in \ker H^\varphi$ and $N^\varphi \subseteq \ker H^\varphi$, from which $\ker H^\varphi = N^\varphi$ and ii) is proved.
iii) Let $v \in V^\varphi_{-\alpha}$ $(0 < \alpha \leq \eta)$ be a primitive vector; then $v \in N^\varphi = \ker H^\varphi$, so that $\mathcal{U}_{\mathbb{k}}.v \subseteq \ker H^\varphi$. Since $\forall x \in \mathcal{U}^-_{\mathbb{k}}$ the map $y \mapsto yx$ is injective, one has that $\dim \mathcal{U}_{\mathbb{k},\alpha-\eta}.v = \mathrm{par}(\eta-\alpha)$, from which the claim follows once one notices that $\mathcal{U}_{\mathbb{k},\alpha-\eta}.v \subseteq V^\varphi_{-\eta}$.
On the other hand if $v \in \ker H^\varphi \cap V^\varphi_{-\eta} \setminus \{0\}$, $\mathcal{U}_{\mathbb{k}}.v \subseteq N^\varphi \subseteq \sum_{\alpha>0} V^\varphi_{-\alpha}$; then, applying the E_i's to v, one finds a primitive homogeneous $w \in \mathcal{U}^+_{\mathbb{k}}.v \cap V^\varphi_{-\alpha} \setminus \{0\}$ with $0 < \alpha \leq \eta$.
iv) follows from the obvious fact that $H^\varphi(v_\varphi, x.v_\varphi) = \pi(x)(\varphi) \ \forall x \in \mathcal{U}_{\mathbb{k}}$: indeed $H^\varphi(f.v_\varphi, \tilde{f}.v_\varphi) = H^\varphi(v_\varphi, \Omega(f)\tilde{f}.v_\varphi) = \pi(\Omega(f)\tilde{f})(\varphi)$. □

Assertion iv) of proposition 1.D4.3 suggests that one could define a "universal" contravariant form H in the following way:

Definition 4.
$H : \mathcal{U}^-_{\mathbb{k}} \times \mathcal{U}^-_{\mathbb{k}} \to \mathcal{U}^0_{\mathbb{k}} = \mathbb{k}[K_0^{\pm 1}, ..., K_n^{\pm 1}]$ is defined by $H(x,y) \doteqdot \pi(\Omega(x)y)$ $\forall x, y \in \mathcal{U}^-_{\mathbb{k}}$.
H_η denotes the restriction of H to $\mathcal{U}^-_{\mathbb{k},-\eta} \times \mathcal{U}^-_{\mathbb{k},-\eta}$.

Remark 5.
H^φ is the evaluation at φ of H, that is $\forall x, y \in \mathcal{U}^-_{\mathbb{k}}$

$$H^\varphi(x.v_\varphi, y.v_\varphi) = H(x,y)(\varphi).$$

Proof: It is an immediate consequence of proposition 1.D4.3. □

Remark 6.
$H|_{\mathcal{U}^-_{\mathbb{k},-\eta} \times \mathcal{U}^-_{\mathbb{k},-\tilde{\eta}}} = 0$ if $\eta \neq \tilde{\eta}$.
Moreover $H(x,y) = \Omega H(y,x) \ \forall x, y \in \mathcal{U}^-_q$.

Proof: The claim follows from proposition 1.D4.3, i) and iv), because π commutes with Ω. □

CHAPTER 2. THE CONTRAVARIANT FORM H, $\det H_\eta$ AND ITS HIGHEST COEFFICIENT.

This chapter is devoted to the investigation of $\det H_\eta$, and in particular of its highest coefficient b_η.
To this aim I first show, using the connections between $\ker H$, the primitive vectors and the pseudo-Casimir operator C, that $\det H_\eta$ is a polynomial in $K_0^{\pm 1},\ldots, K_n^{\pm 1}$ with coefficients in $\mathbb{C}(q)$ which divides a monic polynomial with coefficients in $\mathbb{C}[q,q^{-1}]$.
Then I remark, thanks to the properties of the projection π, that the highest coefficient b_η of $\det H_\eta$ is the determinant of a matrix with coefficients in

$$\mathcal{A} \doteq \mathbb{C}[q,q^{-1},(q^r-q^{-r})^{-1} | 1\leq r\leq \max\{d_0,\ldots,d_n\}],$$

and I try to give this matrix a triangular form: given a linear ordering of $\tilde{R}_+$ there is a condition on the commutation among the "root vectors" such that the corresponding matrix is upper triangular. It is proved that if the ordering of $\tilde{R}_+$ is good (in the sense of definition 1.C2.6), then the commutation of a real root vector with the others satisfies this "triangularity condition", while the commutation among imaginary root vectors does not.
Solving a linear system, I determine then a $\mathbb{C}(q)$-linear transformation of the imaginary root vectors which produces "good" imaginary root vectors; I also study the diagonal terms of the triangular matrix thus obtained; from the study of the relations between b_η and the determinant of this new triangular matrix, I completely determine b_η.

§1. First properties of the contravariant form H and of its determinant.

Remark 1.
In order to work with the contravariant form H I recall that

$$\forall x,y\in\mathcal{U}_q^- \quad H(x,y)\doteq\pi(\Omega(x)y)$$

and that for π the following rules hold:
1) $\pi|_{\mathcal{U}_q^0}=\mathrm{id}_{\mathcal{U}_q^0}$;
2) $\pi(\mathcal{U}_{q,\alpha})=0 \quad \forall\alpha\neq 0$;
3) if $x\in\oplus_{\alpha\in Q_+\setminus\{0\}}\mathcal{U}_{q,-\alpha}$ or $y\in\oplus_{\alpha\in Q_+\setminus\{0\}}\mathcal{U}_{q,\alpha}$ then $\pi(xy)=0$;
4) if $x\in\mathcal{U}_q^0$, $y\in\mathcal{U}_q$ then $\pi(xy)=x\pi(y)$ and $\pi(yx)=\pi(y)x$, that is π is a homomorphism of left and right $\mathcal{U}_q^0$-modules;
5) if $x\in\mathcal{U}_{q,\alpha}$, $y\in\mathcal{U}_{q,\beta}$ and $\lambda\in Q$ then

$$\pi(xK_\lambda y)=q^{(\lambda|\beta)}K_\lambda\pi(xy)=q^{-(\lambda|\alpha)}K_\lambda\pi(xy);$$

6) $\Omega\pi=\pi\Omega$; in particular $\Omega(\pi(\Omega(x)y))=\pi(\Omega(y)x)$, so that

$$\Omega H(x,y) = H(y,x) \ \ \forall x, y \in \mathcal{U}_q^-.$$

□

Definition 2.
$\det H_\eta$ indicates $\det H_\eta(\underline{x},\underline{y})$, where

$$\underline{x} \doteqdot \{x_{\underline{\alpha}} | \underline{\alpha} \in \mathrm{Par}(\eta)\} \ \text{ and } \ \underline{y} \doteqdot \{y_{\underline{\alpha}} | \underline{\alpha} \in \mathrm{Par}(\eta)\}$$

are $\mathbb{C}[q,q^{-1}]$-bases of $\mathcal{U}^-_{\mathbb{C}[q,q^{-1}],-\eta}$ and $H_\eta(\underline{x},\underline{y})_{\underline{\alpha},\underline{\beta}} \doteqdot H(x_{\underline{\alpha}}, y_{\underline{\beta}}) \ \forall \underline{\alpha}, \underline{\beta} \in \mathrm{Par}(\eta)$.
Of course, if thus defined, $\det H_\eta$ depends on the chosen bases, but the following result holds:

Proposition 3.
$\det H_\eta$ is uniquely determined up to units of $\mathbb{C}[q,q^{-1}]$.
Moreover, if $\underline{x} = \{x_{\underline{\alpha}} | \underline{\alpha} \in \mathrm{Par}(\eta)\}$ is a $\mathbb{C}[q,q^{-1}]$-basis of $\mathcal{U}^-_{\mathbb{C}[q,q^{-1}],-\eta}$, then

$$\Omega(\det H(\underline{x},\underline{x})) = \det H(\underline{x},\underline{x}).$$

Proof: Let A and B be the matrices of change of bases from $\underline{x}$ to $\underline{x}'$ and from $\underline{y}$ to $\underline{y}'$ respectively. Then

$$H_\eta(\underline{x}',\underline{y}')_{\underline{\alpha},\underline{\beta}} = H(x'_{\underline{\alpha}}, y'_{\underline{\beta}}) =$$

$$= H\Big(\sum_{\underline{\gamma}\in\mathrm{Par}(\eta)} A_{\underline{\alpha},\underline{\gamma}} x_{\underline{\gamma}}, \sum_{\underline{\gamma}'\in\mathrm{Par}(\eta)} B_{\underline{\beta},\underline{\gamma}'} y_{\underline{\gamma}'}\Big) =$$

$$= \sum_{\underline{\gamma},\underline{\gamma}'\in\mathrm{Par}(\eta)} \Omega(A_{\underline{\alpha},\underline{\gamma}}) H(x_{\underline{\gamma}}, y_{\underline{\gamma}'}) B_{\underline{\beta},\underline{\gamma}'} = (\Omega(A) H_\eta(\underline{x},\underline{y}) {}^t\!B)_{\underline{\alpha},\underline{\beta}}.$$

Hence $\det H_\eta(\underline{x}',\underline{y}') = \det(\Omega(A))\det({}^t\!B)\det H(\underline{x},\underline{y})$.
But A and B (and henceforth $\Omega(A)$ and ${}^t\!B$) are matrices invertible over $\mathbb{C}[q,q^{-1}]$, that is $\det(\Omega(A))\det({}^t\!B)$ is a unit of $\mathbb{C}[q,q^{-1}]$. This proves that $\det H_\eta$ is uniquely determined up to units of $\mathbb{C}[q,q^{-1}]$.
Moreover, since $\Omega(H(x_{\underline{\alpha}}, x_{\underline{\beta}})) = H(x_{\underline{\beta}}, x_{\underline{\alpha}})$, one has that

$$\Omega H_\eta(\underline{x},\underline{x}) = {}^t H_\eta(\underline{x},\underline{x})$$

and

$$\Omega(\det H_\eta(\underline{x},\underline{x})) =$$
$$= \det \Omega H_\eta(\underline{x},\underline{x}) = \det H_\eta(\underline{x},\underline{x}).$$

□

Corollary 4.
There is a unique (up to a nonzero real multiple) Ω-stable polynomial P such that $\det H_\eta = P$.

Proof: The existence of P follows from the above proposition, while the unicity statement follows from the fact that the only Ω-stable units of $\mathbb{C}[q,q^{-1}]$ are the nonzero real numbers. $\square$

Proposition 5.
Let $\alpha\in Q$; then

$$x\in\mathcal{U}^+_{q,\alpha},\quad y\in\mathcal{U}^-_{q,-\alpha},\quad\Rightarrow$$

$$\pi(xy)=\sum_{-\alpha\leq\beta\leq\alpha}a_\beta K_\beta\ \text{ with }\ a_\beta\in\mathbb{C}(q).$$

More generally if $x_i\in\mathcal{U}^+_{q,\gamma_i}$, $y_i\in\mathcal{U}^-_{q,-\tilde{\gamma}_i}$ $\forall i$ and $\gamma=\sum_{i=1}^r\gamma_i$, $\tilde{\gamma}=\sum_{i=1}^r\tilde{\gamma}_i$ then $x_1y_1\cdot\ldots\cdot x_ry_r=\sum_{i=1}^M f_iK_{\xi_i}e_i$ where $\forall i$ $e_i\in\mathcal{U}^+_{q,\gamma'_i}$ and $f_i\in\mathcal{U}^+_{q,-\tilde{\gamma}'_i}$ with $\gamma'_i-\gamma\leq\xi_i\leq\gamma-\gamma'_i$ (or, which is equivalent, $\tilde{\gamma}'_i-\tilde{\gamma}\leq\xi_i\leq\tilde{\gamma}-\tilde{\gamma}'_i$).
Proof: I prove the more general assertion by induction on r.
a) If $r=1$ I proceed by induction on $\mathrm{ht}(\gamma)$: if $\mathrm{ht}(\gamma)$=0 the thesis is obvious, while if $\mathrm{ht}(\gamma)$=1 the thesis follows from the fact that $[E_i,F_j]=\delta_{ij}\dfrac{K_i-K_i^{-1}}{q_i-q_i^{-1}}$, so that $\forall y\in\mathcal{U}^-_{q,-\tilde{\gamma}}$ $[E_i,y]=y_1^{(i)}K_i+y_2^{(i)}K_i^{-1}$ with $y_j^{(i)}\in\mathcal{U}^-_{q,\alpha_i-\tilde{\gamma}}$.
If $\mathrm{ht}(\gamma)>1$ $x=\sum_{i\in I}x_iE_i$ with $x_i\in\mathcal{U}^+_{q,\gamma-\alpha_i}$, hence

$$xy=\sum_{i\in I}x_iE_iy=\sum_{i\in I}x_iyE_i+\sum_{i\in I}x_iy_1^{(i)}K_i+\sum_{i\in I}x_iy_2^{(i)}K_i^{-1}.$$

The thesis then follows by the inductive hypothesis since $\mathrm{ht}(\gamma-\alpha_i)<\mathrm{ht}(\gamma)$.
b) Suppose $r>1$; then

$$(x_1y_1\cdot\ldots\cdot x_{r-1}y_{r-1})(x_ry_r)=\Big(\sum_i f_iK_{\xi_i}e_i\Big)\Big(\sum_j\tilde{f}_jK_{\tilde{\xi}_j}\tilde{e}_j\Big)=$$

$$=\sum_{i,j}f_iK_{\xi_i}e_i\tilde{f}_jK_{\tilde{\xi}_j}\tilde{e}_j=\sum_{i,j,k}f_iK_{\xi_i}\bar{f}_kK_{\bar{\xi}_k}\bar{e}_kK_{\tilde{\xi}_j}\tilde{e}_j$$

where ξ_i, $\tilde{\xi}_j$ and $\bar{\xi}_k$ satisfy the following restrictions:
$\gamma'_i+\gamma_r-\gamma\leq\xi_i\leq\gamma-\gamma_r-\gamma'_i$ with $e_i\in\mathcal{U}^+_{q,\gamma'_i}$,
$\tilde{\gamma}'_j-\gamma_r\leq\tilde{\xi}_j\leq\gamma_r-\tilde{\gamma}'_j$ with $\tilde{e}_j\in\mathcal{U}^+_{q,\tilde{\gamma}'_j}$,
$\bar{\gamma}'_k-\gamma'_i\leq\bar{\xi}_k\leq\gamma'_i-\bar{\gamma}'_k$ with $\bar{e}_k\in\mathcal{U}^+_{q,\bar{\gamma}'_k}$;
then $(\tilde{\gamma}'_j+\bar{\gamma}'_k)-\gamma\leq\xi_i+\tilde{\xi}_j+\bar{\xi}_k\leq\gamma-(\tilde{\gamma}'_j+\bar{\gamma}'_k)$
which implies the claim for $x_1y_1\cdot\ldots\cdot x_ry_r$. $\square$

Corollary 6.
If $f_\gamma\in\mathcal{U}^-_{q,-\gamma}$ and $e\in\mathcal{U}^+_q$, then

$$ef_\gamma=\sum f_\xi K_{\tilde{\xi}}\vec{e}$$

with $\tilde{e}\in\mathcal{U}_q^+$, $f_\xi\in\mathcal{U}_{q,-\xi}^-$, $0\leq\xi\leq\gamma$ and $\xi+\tilde{\xi}\leq\gamma$.

Proof: Of course it is enough to prove the claim for $e=e_{\tilde{\gamma}}\in\mathcal{U}_{q,\tilde{\gamma}}^+$.
Let $e_{\tilde{\gamma}}f_\gamma=\sum f_\xi K_{\tilde{\xi}}\tilde{e}$. Then proposition 2.1.5 implies the claim. □

Corollary 7.

$$\det H_\eta=\sum_{-\mathrm{par}(\eta)\eta\leq\beta\leq\mathrm{par}(\eta)\eta} a_\beta K_\beta \quad\text{with } a_\beta\in\mathbb{C}(q).$$

Proof: Thanks to proposition 2.1.5 the entries of H_η are polynomials of the form $\sum_{-\eta\leq\alpha\leq\eta}a_\alpha K_\alpha$. But H_η is a $\mathrm{par}(\eta)\times\mathrm{par}(\eta)$ matrix, and the thesis follows. □

Recall that I'm looking for

$$\max\{m\in\mathbb{N}|(q-\varepsilon)^m|\det H_\eta\}.$$

The strategy I'll follow will be to prove that

$$\max\{m\in\mathbb{N}|(q-\varepsilon)^m|a_{\mathrm{par}(\eta)\eta}\}=\max\{m\in\mathbb{N}|(q-\varepsilon)^m|\det H_\eta\}$$

and to find

$$\max\{m\in\mathbb{N}|(q-\varepsilon)^m|a_{\mathrm{par}(\eta)\eta}\}.$$

Definition 8.
The highest coefficient of $\det H_\eta$ is the element

$$b_\eta \doteq a_{\mathrm{par}(\eta)\eta}$$

where $\det H_\eta=\sum a_\alpha K_\alpha$; $b_\eta K_\eta^{\mathrm{par}(\eta)}$ is called the highest term of $\det H_\eta$.
More generally, given $x,y\in\mathcal{U}_{q,-\eta}^-$, if $H_\eta(x,y)=\sum a_\alpha K_\alpha$, a_η is said to be the highest coefficient of $H_\eta(x,y)$ and $a_\eta K_\eta$ its highest term.

Proposition 9.
$\exists m>0$ such that $\det H_\eta|P_\eta^m$ over $\mathbb{C}(q)$, where

$$P_\eta=\prod_{0<\alpha\leq\eta}(K_{2\alpha}-q^{(\alpha|\alpha)-2\langle\hat{\rho},\alpha\rangle}).$$

Proof: Let Q be an irreducible factor of $\det H_\eta$; if $\varphi\in(\mathbb{k}^*)^{n+1}$ is such that $Q(\varphi)=0$, also $\det H_\eta(\varphi)=0$, so that, thanks to proposition 1.D4.3, $\exists f\in\mathcal{U}_{q,-\alpha}^-$ such that $f.v_\varphi$ is a primitive vector of V^φ (with $0<\alpha\leq\eta$), hence

$$fv_\varphi=Cfv_\varphi=q^{2\langle\hat{\rho},\alpha\rangle-(\alpha|\alpha)}fK_{2\alpha}Cv_\varphi=$$

$$= q^{2\langle\hat{\rho},\alpha\rangle-(\alpha|\alpha)} f K_{2\alpha} v_\varphi = q^{2\langle\hat{\rho},\alpha\rangle-(\alpha|\alpha)} K_{2\alpha}(\varphi) f v_\varphi;$$

this implies $q^{2\langle\hat{\rho},\alpha\rangle-(\alpha|\alpha)} K_{2\alpha}(\varphi) = 1$, that is φ is a root of $K_{2\alpha} - q^{(\alpha|\alpha)-2\langle\hat{\rho},\alpha\rangle}$, and in particular of P_η.
This means that P_η is zero on the set of the zeros of Q, that is P_η belongs to the ideal generated by Q (because Q is prime, hence radical), so that $Q|P_\eta$.
The claim follows since $\det H_\eta$ is the product of its irreducible factors (with multiplicity). □

Proposition 10.
If $b_\eta \neq 0$ then

$$\frac{\det H_\eta}{b_\eta} \in \mathbb{C}[q, q^{-1}, K_0^{\pm 1}, ..., K_n^{\pm 1}].$$

Proof: Proposition 2.1.9 means that $\exists Q_\eta \in \mathbb{C}(q)[K_0^{\pm 1}, .., K_n^{\pm 1}]$ such that

$$Q_\eta \det H_\eta = P_\eta^m \in \mathbb{C}[q, q^{-1}, K_0^{\pm 1}, ..., K_n^{\pm 1}].$$

Moreover, since $\mathbb{C}(q)$ is the quotient field of $\mathbb{C}[q, q^{-1}]$, Gauss lemma implies that $\exists c \in \mathbb{C}(q)^*$ such that cQ_η, $c^{-1}\det H_\eta \in \mathbb{C}[q, q^{-1}, K_0^{\pm 1}, ..., K_n^{\pm 1}]$; now the leading coefficients of $\det H_\eta$ and Q_η are respectively b_η and b_η^{-1} (because P_η^m is monic), hence cb_η^{-1}, $c^{-1}b_\eta \in \mathbb{C}[q, q^{-1}]$; in particular

$$b_\eta^{-1} \det H_\eta = cb_\eta^{-1} c^{-1} \det H_\eta \in \mathbb{C}[q, q^{-1}, K_0^{\pm 1}, ..., K_n^{\pm 1}].$$

□

§2. How to compute b_η: general strategy for the triangularization of $H_\eta^{\max}$.

I shall now investigate b_η $\forall \eta \in Q_+$.

Definition 1.
Given two families $\underline{x} = \{x_{\underline{\gamma}} | \underline{\gamma} \in \mathrm{Par}(\eta)\}$ and $\underline{y} = \{y_{\underline{\gamma}} | \underline{\gamma} \in \mathrm{Par}(\eta)\}$ of elements of $\mathcal{U}_{q,-\eta}^-$, define $H_\eta^{\max}(\underline{x}, \underline{y})$ as the matrix of the highest coefficients of $H_\eta(\underline{x}, \underline{y})$, that is

$$H(x_{\underline{\alpha}}, y_{\underline{\beta}}) = (H_\eta^{\max}(\underline{x}, \underline{y}))_{\underline{\alpha},\underline{\beta}} K_\eta + \text{lower terms}.$$

Analogously $H_\eta^{\max}$ is the matrix of the highest coefficients of H_η.

Remark 2.
$b_\eta = \det H_\eta^{\max}$
Proof: It immediately follows from proposition 2.1.5. □
The strategy is to transform $H_\eta^{\max}$ into an upper triangular matrix.

Proposition 3.
Suppose there exist:
1) $\mathbb{C}[q,q^{-1}]$-bases $\{x_{\underline{\gamma}}\}_{\underline{\gamma}\in\mathrm{Par}(\eta)}$ and $\{y_{\underline{\gamma}}\}_{\underline{\gamma}\in\mathrm{Par}(\eta)}$ of $\tilde{\mathcal{U}}^+_{\mathbb{C}[q,q^{-1}],\eta}$ and $\tilde{\mathcal{U}}^-_{\mathbb{C}[q,q^{-1}],-\eta}$ respectively;
2) a $\mathbb{C}(q)$-linearly independent set $\{z_{\underline{\gamma}}\}_{\underline{\gamma}\in\mathrm{Par}(\eta)}\subset\mathcal{U}^-_{q,-\eta}$;
3) a linear ordering $\prec$ of $\mathrm{Par}(\eta)$;
satisfying the following property:

$$\forall\underline{\gamma},\underline{\tilde{\gamma}}\in\mathrm{Par}(\eta)\ \ \underline{\gamma}\succ\underline{\tilde{\gamma}}\Rightarrow\pi(x_{\underline{\gamma}}z_{\underline{\tilde{\gamma}}})=\sum_{-\eta\leq\xi<\eta}a_\xi K_\xi;$$

then $b_\eta=(\det A)^{-1}\prod_{\underline{\gamma}\in\mathrm{Par}(\eta)}a_{\underline{\gamma}}$, where $a_{\underline{\gamma}}$ is the coefficient of K_η in $\pi(x_{\underline{\gamma}}z_{\underline{\gamma}})$ and A is the $\mathrm{par}(\eta)\times\mathrm{par}(\eta)$ matrix with coefficients in $\mathbb{C}(q)$ such that $z_{\underline{\gamma}}\doteqdot\sum_{\underline{\tilde{\gamma}}}A_{\underline{\gamma},\underline{\tilde{\gamma}}}y_{\underline{\tilde{\gamma}}}\ \forall\underline{\gamma}\in\mathrm{Par}(\eta)$.

Proof: First remark that $\det A\neq 0$ because the $z_{\underline{\gamma}}$'s are linearly independent. The hypothesis means that the matrix $H^{\max}_\eta(\Omega(\underline{x}),\underline{z})$ is upper triangular; its determinant is then the product of its diagonal terms; but

$$\left(H^{\max}_\eta(\Omega(\underline{x}),\underline{z})\right)_{\underline{\gamma},\underline{\tilde{\gamma}}}=H^{\max}_\eta(\Omega(x_{\underline{\gamma}}),z_{\underline{\tilde{\gamma}}})=$$

$$=H^{\max}_\eta\Big(\Omega(x_{\underline{\gamma}}),\sum_{\underline{\xi}}A_{\underline{\tilde{\gamma}},\underline{\xi}}y_{\underline{\xi}}\Big)=(H^{\max}_\eta(\Omega(\underline{x}),\underline{y})\,{}^t\!A)_{\underline{\gamma},\underline{\tilde{\gamma}}},$$

so that $\det H^{\max}_\eta(\Omega(\underline{x}),\underline{z})=\det(H^{\max}_\eta(\Omega(\underline{x}),\underline{y})\,{}^t\!A)=b_\eta\det A$. □

From now on my effort will be devoted to looking for a situation satisfying the hypotheses of proposition 2.2.3; remark that giving a linearly independent set $\underline{z}=\{z_{\underline{\gamma}}\}_{\underline{\gamma}\in\mathrm{Par}(\eta)}\subset\mathcal{U}^-_{q,-\eta}$ is equivalent to giving an invertible $\mathrm{par}(\eta)\times\mathrm{par}(\eta)$ matrix A with coefficients in $\mathbb{C}(q)$, $\underline{z}$ and A being connected by $\underline{z}=A\underline{y}$.

Definition 4.
Given two elements $x\in\mathcal{U}^+_{q,\alpha}$ and $y\in\mathcal{U}^-_{q,-\beta}$ $(\alpha,\beta\in Q_+)$, (x,y) is said to give no contribution to the highest term if $[x,y]=\sum f_iK_{\xi_i}e_i$ where

$$e_i\in\mathcal{U}^+_{q,\alpha(i)}\setminus\{0\}\text{ and }f_i\in\mathcal{U}^-_{q,-\beta(i)}\setminus\{0\}\ (\alpha(i)-\beta(i)=\alpha-\beta\ \ \forall i)$$

with

$$\xi_i<\alpha-\alpha(i)=\beta-\beta(i)\ (>0)\ \ \forall i.$$

Proposition 5.
Let $x\in\mathcal{U}^+_{q,\alpha}$ and $y\in\mathcal{U}^-_{q,-\beta}$ be such that (x,y) gives no contribution to the highest term and let $\eta\in Q_+$.
Then $\forall x_1,...,x_r,y_1,...,y_r$ such that

$$x_i\in\mathcal{U}^+_{q,\gamma_i},\ y_i\in\mathcal{U}^-_{q,-\tilde{\gamma}_i}\text{ and }\eta=\alpha+\sum_i\gamma_i=\beta+\sum_i\tilde{\gamma}_i,$$

$$\pi(z'[x,y]z'') = \sum_{\xi<\eta} a_\xi K_\xi$$

where $z' = x_1y_1 \cdot \ldots \cdot x_sy_s$ and $z'' = x_{s+1}y_{s+1} \cdot \ldots \cdot x_ry_r$.

Proof: I use the notations of definition 2.2.4:

$$\pi(z'[x,y]z'') = \pi\Big(\sum_j q^{a_j} z' f_j e_j z''\Big) K_{\xi_j} =$$

$$= \sum_{\substack{j,\mu \\ \mu\leq\eta-\alpha+\alpha(j)}} a_\mu^{(j)} K_{\mu+\xi_j} = \sum_{\substack{j,\mu \\ \mu\leq\eta-\alpha+\alpha(j)+\xi_j}} a_{\mu-\xi_j}^{(j)} K_\mu = \sum_{\mu<\eta} \tilde{a}_\mu K_\mu$$

because $\forall j\ \xi_j < \alpha - \alpha(j)$. □

Proposition 6.
Suppose given a linear ordering $\prec$ of $\tilde{R}_+$ and, $\forall\alpha\in\tilde{R}_+$, elements $x_\alpha \in \mathcal{U}^+_{q,p(\alpha)}$ and $y_\alpha \in \mathcal{U}^-_{q,-p(\alpha)}$ such that, $\forall\alpha\succ\beta$, (x_α, y_β) gives no contribution to the highest term and, $\forall\alpha\in\tilde{R}_+$, $[x_\alpha, y_\alpha]\in\mathcal{U}^0_q$.
Then $\forall\eta\in Q_+\ \forall\underline{\gamma},\underline{\tilde{\gamma}}\in\mathrm{Par}(\eta)$

$$\underline{\gamma}\succ\underline{\tilde{\gamma}} \Rightarrow \pi(x(\underline{\gamma},\prec)y(\underline{\tilde{\gamma}},\prec)) = \sum_{-\eta\leq\xi<\eta} a_\xi K_\xi,$$

(recall definition 1.C2.8).

Proof: Induction on η, the cases $\eta = 0$ or $\eta = \alpha_i$ for some $i \in I$ being obvious.
Let $\underline{\gamma} = (\gamma_1, \ldots, \gamma_r)$, $\underline{\tilde{\gamma}} = (\tilde{\gamma}_1, \ldots, \tilde{\gamma}_s) \in \mathrm{Par}(\eta)$ be such that $\underline{\gamma}\succ\underline{\tilde{\gamma}}$. Then if $\underline{\gamma}^* \doteqdot (\gamma_2, \ldots, \gamma_r)$ and $\underline{\tilde{\gamma}}^* \doteqdot (\tilde{\gamma}_2, \ldots, \tilde{\gamma}_s)$ (i.e. $\underline{\gamma} = (\gamma_1, \underline{\gamma}^*)$ and $\underline{\tilde{\gamma}} = (\tilde{\gamma}_1, \underline{\tilde{\gamma}}^*)$),

$$\pi(x(\underline{\gamma},\prec)y(\underline{\tilde{\gamma}},\prec)) = \pi(x(\underline{\gamma},\prec)y_{\tilde{\gamma}_1}y(\underline{\tilde{\gamma}}^*,\prec)) = \pi([x(\underline{\gamma},\prec),y_{\tilde{\gamma}_1}]y(\underline{\tilde{\gamma}}^*,\prec)) =$$

$$= \sum_{\kappa=1}^{r} \pi(x_{\gamma_1}\cdot\ldots\cdot x_{\gamma_{k-1}}[x_{\gamma_k},y_{\tilde{\gamma}_1}]x_{\gamma_{k+1}}\cdot\ldots\cdot x_{\gamma_r}y(\underline{\tilde{\gamma}}^*,\prec)) =$$

$$= \sum_{\substack{k=1 \\ \gamma_k\preceq\tilde{\gamma}_1}}^{r} \pi(x_{\gamma_1}\cdot\ldots\cdot x_{\gamma_{k-1}}[x_{\gamma_k},y_{\tilde{\gamma}_1}]x_{\gamma_{k+1}}\cdot\ldots\cdot x_{\gamma_r}y(\underline{\tilde{\gamma}}^*,\prec)) + \text{lower terms} =$$

$$= \sum_{\substack{k=1 \\ \gamma_k=\tilde{\gamma}_1}}^{r} \pi(x_{\gamma_1}\cdot\ldots\cdot x_{\gamma_{k-1}}[x_{\gamma_k},y_{\tilde{\gamma}_1}]x_{\gamma_{k+1}}\cdot\ldots\cdot x_{\gamma_r}y(\underline{\tilde{\gamma}}^*,\prec)) + \text{lower terms} =$$

$$= \sum_{\substack{k=1 \\ \gamma_k=\tilde{\gamma}_1}}^{r} \pi(x_{\gamma_2}\cdot\ldots\cdot x_{\gamma_r}y(\underline{\tilde{\gamma}}^*,\prec))a_{\tilde{\gamma}_1}K_{\tilde{\gamma}_1} + \text{lower terms} =$$

$$= a\pi(x(\underline{\gamma}^*,\prec)y(\underline{\tilde{\gamma}}^*,\prec))K_{\tilde{\gamma}_1} + \text{lower terms}$$

where "lower terms" stands for "terms of lower degree than K_η" and $a=0$ if $\gamma_1 \succ \tilde{\gamma}_1$; if $\gamma_1 = \tilde{\gamma}_1$ then $\underline{\gamma}^* \succ \underline{\tilde{\gamma}}^*$ and both belong to $\mathrm{Par}(\eta - p(\gamma_1))$, so that the claim follows by the inductive hypothesis. □

§3. The real root vectors: commutation and contribution to the highest term.

From now on my effo.t will be in looking for x_β, y_β ($\beta \in \tilde{R}_+$) satisfying the properties of proposition 2.2.6.
Of course, since $\forall \alpha \in \tilde{R}_+$ I have elements $E_\alpha \in \mathcal{U}^+_{q,p(\alpha)}$, and $F_\alpha \in \mathcal{U}^-_{q,-p(\alpha)}$, I start the investigation from these vectors.

Proposition 1.

$$\forall \alpha \in R^{\mathrm{re}}_+ \quad [E_\alpha, F_\alpha] = \frac{K_\alpha - K_\alpha^{-1}}{q_\alpha - q_\alpha^{-1}}$$

where $q_\alpha \doteqdot q_i$ if $\alpha = w(\alpha_i)$ for some $w \in W$.
Proof: There exists $T \in \mathcal{B}$ such that $E_\alpha = T(E_i)$ and $F_\alpha = T(F_i)$; of course α and α_i are conjugate under W, so that $q_\alpha = q_i$. Then

$$[E_\alpha, F_\alpha] = T[E_i, F_i] = T\Big(\frac{K_i - K_i^{-1}}{q_i - q_i^{-1}}\Big) = \frac{K_\alpha - K_\alpha^{-1}}{q_\alpha - q_\alpha^{-1}}.$$

□

In what follows, I want to prove that $\{E_\alpha | \alpha \in \tilde{R}_+\}$, $\{F_\alpha | \alpha \in \tilde{R}_+\}$ are "near" to satisfying the hypotheses of proposition 2.2.6 with respect to a good ordering $\prec$ of $\tilde{R}_+$.
To this aim I proceed by steps, proving what I need of the commutation formulas between the root vectors.

Proposition 2.
Let $\prec$ be a good ordering of $\tilde{R}_+$ and let $h, k \in \mathbb{Z}$ be such that $\beta_h \succ \beta_k$; then $(E_{\beta_h}, F_{\beta_k})$ gives no contribution to the highest ter:n.
Proof: Recall that $\beta_h \succ \beta_k$ means either $h > k > 0$ or $0 \geq h > k$ or $h \leq 0 < k$.
If $k > 0$, let w and T be defined by

$$w \doteqdot s_{i_1} \cdot \ldots \cdot s_{i_k}, \quad T \doteqdot T_w.$$

Remark the following facts:
1) $T^{-1}(E_{\beta_h}) \in \mathcal{U}_q^+$: indeed

$$T^{-1}(E_{\beta_h}) = \begin{cases} T_{i_k}^{-1} \cdot \ldots \cdot T_{i_1}^{-1} T_{i_1} \cdot \ldots \cdot T_{i_{h-1}}(E_{i_h}) & \text{if } h > 0 \\ T_{i_k}^{-1} \cdot \ldots \cdot T_{i_1}^{-1} T_{i_0}^{-1} \cdot \ldots \cdot T_{i_{h+1}}^{-1}(E_{i_h}) & \text{if } h \leq 0 \end{cases}$$

$$= \begin{cases} T_{i_{k+1}} \cdot \ldots \cdot T_{i_{h-1}}(E_{i_h}) & \text{if } h > 0 \\ T_{i_k}^{-1} \cdot \ldots \cdot T_{i_1}^{-1} T_{i_0}^{-1} \cdot \ldots \cdot T_{i_{h+1}}^{-1}(E_{i_h}) & \text{if } h \leq 0 \end{cases}$$

which is an element of $\mathcal{U}_q^+$ because $s_{i_{k+1}} \cdot \ldots \cdot s_{i_h}$ $(h > 0)$ and $s_{i_k} \cdot \ldots \cdot s_{i_h}$ $(h \leq 0)$ are reduced expressions.
2) $T^{-1}(F_{\beta_k}) = -E_{i_k} K_{i_k}$: indeed

$$T^{-1}(F_{\beta_k}) = T_{i_k}^{-1} \cdot \ldots \cdot T_{i_1}^{-1} T_{i_1} \cdot \ldots \cdot T_{i_{k-1}}(F_{i_k}) = T_{i_k}^{-1}(F_{i_k}) = -E_{i_k} K_{i_k}.$$

3) $T(K_{i_k}) = K_{-\beta_k}$: indeed

$$s_{i_1} \cdot \ldots \cdot s_{i_k}(\alpha_{i_k}) = -s_{i_1} \cdot \ldots \cdot s_{i_{k-1}}(\alpha_{i_k}) = -\beta_k.$$

Then $[E_{\beta_h}, F_{\beta_k}] = T[T^{-1}(E_{\beta_h}), T^{-1}(F_{\beta_k})] = T(EK_{i_k}) = T(E)K_{-\beta_k}$ with $E \in \mathcal{U}_q^+$.
Now, using corollary 1.C1.12 and corollary 2.1.6,

$$[E_{\beta_h}, F_{\beta_k}] = \sum e f_\gamma K_{\gamma - \beta_k} = \sum f_\xi K_{\gamma - \beta_k + \tilde{\xi}} \tilde{e}$$

with $0 \leq \xi \leq \gamma < \beta_k$ and $\xi + \tilde{\xi} \leq \gamma$. Hence $\gamma - \beta_k + \tilde{\xi} < \beta_k - \xi$ because $\gamma + \xi + \tilde{\xi} \leq 2\gamma < 2\beta_k$. This implies the assertion for $k > 0$.
Analogously, if $k \leq 0$ let w and T be defined by

$$w \doteqdot s_{i_h} \cdot \ldots \cdot s_{i_0}, \quad T \doteqdot T_w^{-1}.$$

Remark the following facts:
1) $T^{-1}(F_{\beta_k}) \in \mathcal{U}_q^-$: indeed

$$T^{-1}(F_{\beta_k}) = T_{i_h} \cdot \ldots \cdot T_{i_0} T_{i_0}^{-1} \cdot \ldots \cdot T_{i_{k+1}}^{-1}(F_{i_k}) = T_{i_{h-1}}^{-1} \cdot \ldots \cdot T_{i_{k+1}}^{-1}(F_{i_k})$$

which is an element of $\mathcal{U}_q^-$ because $s_{i_{h-1}} \cdot \ldots \cdot s_{i_k}$ is a reduced expression.
2) $T^{-1}(E_{\beta_h}) = -F_{i_h} K_{i_h}$: indeed

$$T^{-1}(E_{\beta_h}) = T_{i_h} \cdot \ldots \cdot T_{i_0} T_{i_0}^{-1} \cdot \ldots \cdot T_{i_{h+1}}^{-1}(E_{i_h}) = T_{i_h}(E_{i_h}) = -F_{i_h} K_{i_h}.$$

3) $T(K_{i_h}) = K_{-\beta_h}$: indeed

$$s_{i_0} \cdot \ldots \cdot s_{i_h}(\alpha_{i_h}) = -s_{i_0} \cdot \ldots \cdot s_{i_{h+1}}(\alpha_{i_h}) = -\beta_h.$$

Then $[E_{\beta_h}, F_{\beta_k}] = T[T^{-1}(E_{\beta_h}), T^{-1}(F_{\beta_k})] = T(FK_{i_h}) = T(F)K_{-\beta_h}$ with $F \in \mathcal{U}_q^-$.
Now, thanks to corollary 1.C1.12,

$$[E_{\beta_h}, F_{\beta_k}] = \sum f K_{\gamma - \beta_h} e_\gamma$$

with $f \in \mathcal{U}_q^-$, $e_\gamma \in \mathcal{U}_{q,\gamma}^+$ and the claim follows since $\gamma < \beta_h$, so that $\gamma - \beta_h < \beta_h - \gamma$. □

Proposition 3.
1) If $r \leq 0$ and $X \in \mathcal{U}_{q,-\alpha}^-$ is $T_{2\rho^\cdot}$-stable then (E_{β_r}, X) gives no contribution to the highest term;
2) if $r > 0$ and $X \in \mathcal{U}_{q,\alpha}^+$ is $T_{2\rho^\cdot}$-stable then (X, F_{β_r}) gives no contribution to the highest term.

Proof: 1) Let $T \doteq T_{2\rho^\cdot}^s$ where s is such that $Ns < r$. Since X is T-stable, one has that

$$[E_{\beta_r}, X] = T[T^{-1}(E_{\beta_r}), X].$$

But, since $\underline{i}$ is cyclic of period N,

$$T^{-1}(E_{\beta_r}) = T_{i_1} \cdot \ldots \cdot T_{i_{-Ns}} T_{i_0}^{-1} \cdot \ldots \cdot T_{i_{r+1}}^{-1}(E_{i_r}) =$$

$$= T_{i_1} \cdot \ldots \cdot T_{i_{-Ns}} T_{i_{-Ns}}^{-1} \cdot \ldots \cdot T_{i_{r-Ns+1}}^{-1}(E_{i_{r-Ns}}) =$$

$$= T_{i_1} \cdot \ldots \cdot T_{i_{r-Ns}}(E_{i_{r-Ns}}) = -T_{i_1} \cdot \ldots \cdot T_{i_{r-Ns-1}}(F_{i_{r-Ns}} K_{i_{r-Ns}}) =$$

$$= -F_{\beta_{r-Ns}} K_{\beta_{r-Ns}}.$$

Hence

$$[E_{\beta_r}, X] = T(FK_{\beta_{r-Ns}}) = T(F)K_{-\beta_r}$$

(with $F \in \mathcal{U}_q^-$) because

$$T(K_{\beta_{r-Ns}}) = T_{i_{-Ns}}^{-1} \cdot \ldots \cdot T_{i_1}^{-1} T_{i_1} \cdot \ldots \cdot T_{i_{r-Ns-1}}(K_{i_{r-Ns}}) =$$

$$= T_{i_{-Ns}}^{-1} \cdot \ldots \cdot T_{i_{r-Ns}}^{-1}(K_{i_{r-Ns}}) = T_{i_0}^{-1} \cdot \ldots \cdot T_{i_{r+1}}^{-1}(K_{i_r}^{-1}) = K_{-\beta_r}.$$

This implies, thanks to corollary 1.C1.12, that

$$[E_{\beta_r}, X] = \sum_{0 \leq \gamma < \beta_r} f K_{\gamma - \beta_r} e_\gamma$$

which is the claim because $\gamma - \beta_r < \beta_r - \gamma$.
2) Let $T \doteq T_{2\rho^\cdot}^s$ where $Ns \geq r$. Since X is T-stable, one has that

$$[X, F_{\beta_r}] = T[X, T^{-1}(F_{\beta_r})].$$

But

$$T^{-1}(F_{\beta_r}) = T_{i_{Ns}}^{-1} \cdot \ldots \cdot T_{i_1}^{-1} T_{i_1} \cdot \ldots \cdot T_{i_{r-1}}(F_{i_r}) = T_{i_{Ns}}^{-1} \cdot \ldots \cdot T_{i_r}^{-1}(F_{i_r}) =$$

$$= -T_{i_{Ns}}^{-1} \cdot \ldots \cdot T_{i_{r+1}}^{-1}(E_{i_r} K_{i_r}) = -T_{i_0}^{-1} \cdot \ldots \cdot T_{i_{r+1-Ns}}^{-1}(E_{i_{r-Ns}} K_{i_{r-Ns}}) =$$

$$= -E_{\beta_{r-N_s}} K_{\beta_{r-N_s}}.$$

Hence

$$[X, F_{\beta_r}] = T(EK_{\beta_{r-N_s}}) = T(E)K_{-\beta_r}$$

(with $E \in \mathcal{U}_q^+$) because

$$T(K_{\beta_{r-N_s}}) = T_{i_1} \cdot \ldots \cdot T_{i_{N_s}} T_{i_0}^{-1} \cdot \ldots \cdot T_{i_{r-N_s+1}}^{-1}(K_{i_{r-N_s}}) =$$

$$= T_{i_1} \cdot \ldots \cdot T_{i_{N_s}} T_{i_{N_s}}^{-1} \cdot \ldots \cdot T_{i_{r+1}}^{-1}(K_{i_r}) = T_{i_1} \cdot \ldots \cdot T_{i_r}(K_{i_r}) =$$

$$= T_{i_1} \cdot \ldots \cdot T_{i_{r-1}}(K_{i_r}^{-1}) = K_{-\beta_r}.$$

This implies, thanks to corollary 1.C1.12 and corollary 2.1.6, that

$$[X, F_{\beta_r}] = \sum_{0 \le \gamma < \beta_r} e f_\gamma K_{\gamma-\beta_r} = \sum f_\xi K_{\gamma-\beta_r+\tilde{\xi}} \tilde{e}$$

with $0 \le \xi \le \gamma < \beta_r$ and $\tilde{\xi} \le \gamma - \xi$, so that $\gamma - \beta_r + \tilde{\xi} < \tilde{\xi} \le \gamma - \xi < \beta_r - \xi$, which is the claim. □

Corollary 4.
Let $\prec$ be a good ordering of $\tilde{R}_+$; then
1) If $\alpha, \beta \in \tilde{R}_+$ are not both in $R_+^{\text{im}} \times I_0$, then $\alpha \succ \beta \Rightarrow (E_\alpha, F_\beta)$ gives no contribution to the highest term;
2) if $\alpha \in R_+^{\text{re}}$ and $X \in \mathcal{U}_{q,\text{im}}^+$ then either $\alpha = \beta_r$ with $r > 0$ and (X, F_α) gives no contribution to the highest term or $\alpha = \beta_r$ with $r \le 0$ and $(E_\alpha, \Omega(X))$ gives no contribution to the highest term.

Proof: 1) follows from proposition 2.3.2 and proposition 2.3.3, since the imaginary root vectors are T_ρ-stable and $\beta_r \prec (m\delta, i) \prec \beta_s$ $\forall r, m > 0$, $\forall s \le 0$, $\forall i \in I_0$ and for every good ordering $\prec$; 2) immediately follows from proposition 2.3.3 since $E_{(m\delta,i)}$ is T_ρ-stable $\forall m > 0$, $\forall i \in I$. □

§4. Commutation rules for the imaginary root vectors.

Proposition 1.
$i, j \in I_0$ such that $a_{i,j} = 0 \Rightarrow \mathcal{U}_q(i)$ commutes with $\mathcal{U}_q(j)$.

Proof: Let i, j be such that $a_{i,j} = 0$.
Notice that K_i and $K_{\delta-\alpha_i}$ commute with $\mathcal{U}_q(j)$ (and K_j and $K_{\delta-\alpha_j}$ commute with $\mathcal{U}_q(i)$).
Moreover $F_i = \Omega(E_i)$,

$$E_{\delta-\alpha_i} = T_{\omega_i} T_i^{-1}(E_i) = T_i T_{\omega_i}^{-1}(E_i), \quad F_{\delta-\alpha_i} = T_i T_{\omega_i}^{-1}(F_i)$$

and $\mathcal{U}_q(j)$ is Ω-stable and $T_{\omega_i} T_i^{-1}$-stable, so that it is enough to prove that E_i commutes with $\mathcal{U}_q(j)$.

But $T_{\omega_j^-}T_j^{-1}(E_i) = E_i$, hence the only thing to prove is that $[E_i, E_j] = 0$, $[E_i, F_j] = 0$, which is obviously true. □

Corollary 2.
$i, j \in I_0$ such that $a_{i,j} = 0 \Rightarrow \forall r, s > 0$

$$[E_{(r\delta,i)}, F_{(s\delta,j)}] = 0.$$

□

Lemma 3.
$\forall i \neq j \in I_0$ $T_{\omega_j^-}(E_iE_j - q_i^{a_{i,j}}E_jE_i) = T_{\omega_i^-}(E_jE_i - q_i^{a_{i,j}}E_iE_j)$.
Proof: See Beck [1]. □

Remark 4.
Recall that $\forall m > 0$, $\forall i \in I_0$ and $\forall r \in \mathbb{Z}$

$$\tilde{E}_{(m\delta,i)} = K_{r\delta-\alpha_i}[T_{\omega_i^-}^r(F_i), T_{\omega_i^-}^{r-m}(E_i)];$$

indeed if $r = m$ this is the definition of $\tilde{E}_{(m\delta,i)}$, and in general $\forall s \in \mathbb{Z}$

$$\tilde{E}_{(m\delta,i)} = T_{\omega_i^-}^s(\tilde{E}_{(m\delta,i)}) = T_{\omega_i^-}^s(K_{m\delta-\alpha_i}[T_{\omega_i^-}^m(F_i), E_i]) =$$

$$= K_{(m+s)\delta-\alpha_i}[T_{\omega_i^-}^{m+s}(F_i), T_{\omega_i^-}^s(E_i)].$$

□

For sake of homogeneity of notations I set the following definition:

Definition 5.
$\forall i \in I_0$ I set

$$\tilde{E}_{(0,i)} \doteq K_i^{-1}[F_i, E_i] = \frac{K_i^{-2} - 1}{q_i - q_i^{-1}}.$$

Remark 6.
Let $x \in \mathcal{U}_q$ be such that $K_i x = q_i^{-a_{i,j}} x K_i$; then $\forall s \in \mathbb{Z}$

$$q_i^{a_{i,j}} x\tilde{E}_{(0,i)} - q_i^{-a_{i,j}}\tilde{E}_{(0,i)}x = q_i^{a_{i,j}} xT_{\omega_i^-}^s(\tilde{E}_{(0,i)}) - q_i^{-a_{i,j}}T_{\omega_i^-}^s(\tilde{E}_{(0,i)})x.$$

Proof: $T_{\omega_i^-}^s(\tilde{E}_{(0,i)}) = \dfrac{K_{s\delta-\alpha_i}^2 - 1}{q_i - q_i^{-1}}$; thus

$$q_i^{a_{i,j}} xT_{\omega_i^-}^s(\tilde{E}_{(0,i)}) - q_i^{-a_{i,j}}T_{\omega_i^-}^s(\tilde{E}_{(0,i)})x =$$

$$= q_i^{a_{i,j}} x\frac{K_{s\delta-\alpha_i}^2 - 1}{q_i - q_i^{-1}} - q_i^{-a_{i,j}}\frac{K_{s\delta-\alpha_i}^2 - 1}{q_i - q_i^{-1}}x =$$

$$= \frac{q_i^{-a_{i,j}} K_{s\delta-\alpha_i}^2 x}{q_i - q_i^{-1}} - \frac{q_i^{a_{i,j}} - q_i^{-a_{i,j}}}{q_i - q_i^{-1}} x - \frac{q_i^{-a_{i,j}} K_{s\delta-\alpha_i}^2 x}{q_i - q_i^{-1}} = -[a_{i,j}]_{q_i} x$$

which is independent of s. □

Lemma 7.
$\forall i, j \in I_0$ such that $a_{i,j} < 0$, $\forall m > 0$ and $\forall r \in \mathbb{Z}$

$$[T_{\omega_j^\vee}^{-r}(E_j), \tilde{E}_{(m\delta,i)}] = (q_i^{a_{i,j}} - q_i^{-a_{i,j}}) \tilde{E}_{((m-1)\delta,i)} T_{\omega_j^\vee}^{-r-1}(E_j) +$$

$$-q_i^{-a_{i,j}} [T_{\omega_j^\vee}^{-r-1}(E_j), \tilde{E}_{((m-1)\delta,i)}].$$

Proof: Notice first that if $m > 0$ and $i, j \in I_0$ are such that $a_{i,j} < 0$ and

$$[E_j, \tilde{E}_{(m\delta,i)}] =$$

$$= (q_i^{a_{i,j}} - q_i^{-a_{i,j}}) \tilde{E}_{((m-1)\delta,i)} T_{\omega_j^\vee}^{-1}(E_j) - q_i^{-a_{i,j}} [T_{\omega_j^\vee}^{-1}(E_j), \tilde{E}_{((m-1)\delta,i)}],$$

then, applying $T_{\omega_j^\vee}^{-r}$ and remarking that $\tilde{E}_{(s\delta,i)}$ is $T_{\omega_j^\vee}$-stable $\forall s \geq 0$, it is also true that $\forall r \in \mathbb{Z}$

$$[T_{\omega_j^\vee}^{-r}(E_j), \tilde{E}_{(m\delta,i)}] = (q_i^{a_{i,j}} - q_i^{-a_{i,j}}) \tilde{E}_{((m-1)\delta,i)} T_{\omega_j^\vee}^{-r-1}(E_j) +$$

$$-q_i^{-a_{i,j}} [T_{\omega_j^\vee}^{-r-1}(E_j), \tilde{E}_{((m-1)\delta,i)}].$$

Hence it is enough to prove the claim for $r = 0$:

$$[E_j, \tilde{E}_{(m\delta,i)}] = [E_j, K_i^{-1}[F_i, T_{\omega_i^\vee}^{-m}(E_i)]] =$$

$$= K_i^{-1}\big(q_i^{a_{i,j}} E_j F_i T_{\omega_i^\vee}^{-m}(E_i) - q_i^{a_{i,j}} E_j T_{\omega_i^\vee}^{-m}(E_i) F_i +$$

$$-F_i T_{\omega_i^\vee}^{-m}(E_i) E_j + T_{\omega_i^\vee}^{-m}(E_i) F_i E_j\big) =$$

$$= K_i^{-1}[F_i, T_{\omega_i^\vee}^{-m}(q_i^{a_{i,j}} E_j E_i - E_i E_j)] =$$

$$= K_i^{-1}[F_i, T_{\omega_i^\vee}^{1-m} T_{\omega_j^\vee}^{-1}(q_i^{a_{i,j}} E_i E_j - E_j E_i)] =$$

$$= T_{\omega_j^\vee}^{-1}(K_i^{-1}[F_i, q_i^{a_{i,j}} T_{\omega_i^\vee}^{1-m}(E_i) E_j - E_j T_{\omega_i^\vee}^{1-m}(E_i)]) =$$

$$= T_{\omega_j^\vee}^{-1}(K_i^{-1}(q_i^{a_{i,j}} [F_i, T_{\omega_i^\vee}^{1-m}(E_i)] E_j - E_j [F_i, T_{\omega_i^\vee}^{1-m}(E_i)])) =$$

$$= T_{\omega_j^\vee}^{-1}(-q_i^{-a_{i,j}} [E_j, \tilde{E}_{((m-1)\delta,i)}] + (q_i^{a_{i,j}} - q_i^{-a_{i,j}}) \tilde{E}_{((m-1)\delta,i)} E_j) =$$

$$= -q_i^{-a_{i,j}} [T_{\omega_j^\vee}^{-1}(E_j), \tilde{E}_{((m-1)\delta,i)}] + (q_i^{a_{i,j}} - q_i^{-a_{i,j}}) \tilde{E}_{((m-1)\delta,i)}] T_{\omega_j^\vee}^{-1}(E_j).$$

□

Lemma 8.
$\forall i, j \in I_0$ such that $a_{i,j} < 0$, $\forall m > 0$ and $\forall r \in \mathbb{Z}$

$$[\tilde{E}_{(m\delta,i)}, T_{\omega_j^\vee}^r(-K_j^{-1}F_j)] = (q_i^{a_{i,j}} - q_i^{-a_{i,j}})T_{\omega_j^\vee}^{r+1}(-K_j^{-1}F_j)\tilde{E}_{((m-1)\delta,i)} +$$

$$-q_i^{-a_{i,j}}[\tilde{E}_{((m-1)\delta,i)}, T_{\omega_j^\vee}^{r+1}(-K_j^{-1}F_j)].$$

Proof: Exactly as in lemma 2.4.7 it is enough to prove the claim for $r = 0$. Moreover lemma 2.4.3 implies, applying Ω, that

$$F_iF_j - q_i^{-a_{i,j}}F_jF_i = T_{\omega_i^\vee}^{-1}T_{\omega_j^\vee}(F_jF_i - q_i^{-a_{i,j}}F_iF_j).$$

Then

$$[\tilde{E}_{(m\delta,i)}, F_j] = [K_i^{-1}[F_i, T_{\omega_i^\vee}^{-m}(E_i)], F_j] =$$

$$= K_i^{-1}([F_iF_j - q_i^{-a_{i,j}}F_jF_i, T_{\omega_i^\vee}^{-m}(E_i)]) =$$

$$= K_i^{-1}T_{\omega_j^\vee}([F_jT_{\omega_i^\vee}^{-1}(F_i) - q_i^{-a_{i,j}}T_{\omega_i^\vee}^{-1}(F_i)F_. \ T_{\omega_i^\vee}^{-m}(E_i)]) =$$

$$= T_{\omega_j^\vee}(K_i^{-1}(F_j[T_{\omega_i^\vee}^{-1}(F_i), T_{\omega_i^\vee}^{-m}(E_i)] - q_i^{-a_{i,j}}[T_{\omega_i^\vee}^{-1}(F_i), T_{\omega_i^\vee}^{-m}(E_i)]F_j)) =$$

$$= T_{\omega_j^\vee}T_{\omega_i^\vee}^{-1}(K_{\delta-\alpha_i}(F_j[F_i, T_{\omega_i^\vee}^{1-m}(E_i)] - q_i^{-a_{i,j}}[F_i, T_{\omega_i^\vee}^{1-m}(E_i)]F_j)) =$$

$$= K_\delta T_{\omega_j^\vee}T_{\omega_i^\vee}^{-1}(q_i^{a_{i,j}}F_j\tilde{E}_{((m-1)\delta,i)} - q_i^{-a_{i,j}}\tilde{E}_{((m-1)\delta,i)}F_j) =$$

$$= K_\delta T_{\omega_j^\vee}T_{\omega_i^\vee}^{-1}(q_i^{a_{i,j}}F_jT_{\omega_i^\vee}(\tilde{E}_{((m-1)\delta,i)}) - q_i^{-a_{i,j}}T_{\omega_i^\vee}(\tilde{E}_{((m-1)\delta,i)})F_j) =$$

$$= K_\delta((q_i^{a_{i,j}} - q_i^{-a_{i,j}})T_{\omega_j^\vee}(F_j)\tilde{E}_{((m-1)\delta,i)} - q_i^{-a_{i,j}}[\tilde{E}_{((m-1)\delta,i)}, T_{\omega_j^\vee}(F_j)])$$

because $K_iF_j = q_i^{-a_{i,j}}F_jK_i$.
The thesis follows noticing that $T_{\omega_j^\vee}(K_j^{-1}) = K_\delta K_j^{-1}$. $\qquad\square$

Lemma 9.
Let A be a $\mathbb{C}(q)$-algebra and let T, $\tilde{S}_m^+$, $\tilde{S}_m^-$ $(m > 0)$ be commuting operators on A, with T invertible. Let $x \in A$ be an element such that

$$(\tilde{S}_1^+ - \tilde{S}_1^-)(x) = -[a]_v T(x)$$

and, $\forall m > 1$,

$$(\tilde{S}_m^+ - \tilde{S}_m^-)(x) = (v^a - v^{-a})\tilde{S}_{m-1}^- T(x) - v^{-a}(\tilde{S}_{m-1}^+ - \tilde{S}_{m-1}^-)T(x),$$

where $v \in \mathbb{C}(q)^*$, $a \in \mathbb{Z}$ and $[a]_v = v^{-a+1} + v^{-a+3} + \dots + v^{a-1}$.
Then $\forall m > 0$, $\forall r \in \mathbb{Z}$

$$(\tilde{S}_m^+ - \tilde{S}_m^-)T^r(x) =$$

$$= -[a]_v \left(\sum_{s=1}^{m-1} (-1)^s v^{-(s-1)a}(v-v^{-1})\tilde{S}^-_{m-s}T^{s+r} - (-1)^m v^{-(m-1)a}T^{m+r}\right)(x).$$

Proof: I prove the claim by induction on m.
If $m=1$ the claim is true applying T^r to $(\tilde{S}^+_1 - \tilde{S}^-_1)(x) = -[a]_v T(x)$.
If $m>1$, $(\tilde{S}^+_m - \tilde{S}^-_m)(x) = (v^a - v^{-a})\tilde{S}^-_{m-1}T(x) - v^{-a}(\tilde{S}^+_{m-1} - \tilde{S}^-_{m-1})T(x) =$

$$= (v^a - v^{-a})\tilde{S}^-_{m-1}T(x) + [a]_v v^{-a} T\left(\sum_{s=1}^{m-2} (-1)^s v^{-(s-1)a}(v-v^{-1})\tilde{S}^-_{m-s-1}T^s + \right.$$

$$\left. -(-1)^{m-1}v^{-(m-2)a}T^{m-1}\right)(x) =$$

$$= (v-v^{-1})[a]_v \tilde{S}^-_{m-1}T(x) - [a]_v \left(\sum_{s=2}^{m-1} (-1)^s v^{-(s-1)a}(v-v^{-1})\tilde{S}^-_{m-s}T^s(x) + \right.$$

$$\left. -(-1)^m v^{-(m-1)a}T^m(x)\right) =$$

$$= -[a]_v \left(\sum_{s=1}^{m-1} (-1)^s v^{-(s-1)a}(v-v^{-1})\tilde{S}^-_{m-s}T^s(x) + \right.$$

$$\left. -(-1)^m v^{-(m-1)a}T^m(x)\right),$$

and it is enough to apply T^r to get the claim. □

Proposition 10.
$\forall i,j \in I_0$ with $a_{i,j} < 0$ $\forall m > 0$ $\forall r \in \mathbb{Z}$

$$[T^{-r}_{\omega_j^\vee}(E_j), \tilde{E}_{(m\delta,i)}] =$$

$$= -[a_{i,j}]_{q_i} \left(\sum_{s=1}^{m-1} (-1)^s q_i^{-(s-1)a_{i,j}}(q_i - q_i^{-1})\tilde{E}_{((m-s)\delta,i)}T^{-r-s}_{\omega_j^\vee}(E_j) + \right.$$

$$\left. -(-1)^m q_i^{-(m-1)a_{i,j}} T^{-r-m}_{\omega_j^\vee}(E_j)\right)$$

and

$$[\tilde{E}_{(m\delta,i)}, T^r_{\omega_j^\vee}(-K_j^{-1}F_j)] =$$

$$= -[a_{i,j}]_{q_i} \left(\sum_{s=1}^{m-1} (-1)^s q_i^{-(s-1)a_{i,j}}(q_i - q_i^{-1})T^{r+s}_{\omega_j^\vee}(-K_j^{-1}F_j)\tilde{E}_{((m-s)\delta,i)} + \right.$$

$$-(-1)^m q_i^{-(m-1)a_{i,j}} T_{\omega_j^\vee}^{r+m}(-K_j^{-1}F_j)\Big).$$

Proof: Let $i,j\in I_0$ with $a_{i,j}<0$ be fixed.
$\forall m>0$ I introduce the following operators on $\mathcal{U}_q$:

$$\tilde{S}_m^+(x) \doteq x\tilde{E}_{(m\delta,i)} \quad \forall x\in\mathcal{U}_q,$$

$$\tilde{S}_m^-(x) \doteq \tilde{E}_{(m\delta,i)}x \quad \forall x\in\mathcal{U}_q;$$

moreover I set $T\doteq T_{\omega_j^\vee}^{-1}$. These are of course commuting operators on $\mathcal{U}_q$, because $T(\tilde{E}_{(m\delta,i)})=\tilde{E}_{(m\delta,i)}$ $\forall m>0$, $\forall i\in I_0$.
Remark that lemma 2.4.7 implies that $\forall m>1$

$$(\tilde{S}_m^+-\tilde{S}_m^-)(E_j)=(q_i^{a_{i,j}}-q_i^{-a_{i,j}})\tilde{S}_{m-1}^-T(E_j)-q_i^{-a_{i,j}}(\tilde{S}_{m-1}^+-\tilde{S}_{m-1}^-)T(E_j)$$

while lemma 2.4.8 implies that $\forall m>1$

$$(\tilde{S}_m^--\tilde{S}_m^+)(-K_j^{-1}F_j)=(q_i^{a_{i,j}}-q_i^{-a_{i,j}})\tilde{S}_{m-1}^+T^{-1}(-K_j^{-1}F_j)+$$

$$-q_i^{-a_{i,j}}(\tilde{S}_{m-1}^--\tilde{S}_{m-1}^+)T^{-1}(-K_j^{-1}F_j).$$

Then, thanks to lemma 2.4.9, it is enough to prove that

$$(\tilde{S}_1^+-\tilde{S}_1^-)(E_j)=-[a_{i,j}]_{q_i}T(E_j)$$

and

$$(\tilde{S}_1^--\tilde{S}_1^+)(-K_j^{-1}F_j)=-[a_{i,j}]_{q_i}T^{-1}(-K_j^{-1}F_j).$$

Now, lemmas 2.4.7 and 2.4.8 imply that

$$(\tilde{S}_1^+-\tilde{S}_1^-+[a_{i,j}]_{q_i}T)(E_j)=$$

$$=q_i^{a_{i,j}}\frac{K_i^{-2}-1}{q_i-q_i^{-1}}T(E_j)-q_i^{-a_{i,j}}T(E_j)\frac{K_i^{-2}-1}{q_i-q_i^{-1}}+[a_{i,j}]_{q_i}T(E_j)=$$

$$=\frac{q_i^{-a_{i,j}}-q_i^{a_{i,j}}}{q_i-q_i^{-1}}T(E_j)+[a_{i,j}]_{q_i}T(E_j)=0,$$

and

$$(\tilde{S}_1^--\tilde{S}_1^++[a_{i,j}]_{q_i}T^{-1})(-K_j^{-1}F_j)=$$

$$=q_i^{a_{i,j}}T^{-1}(-K_j^{-1}F_j)\frac{K_i^{-2}-1}{q_i-q_i^{-1}}+$$

$$-q_i^{-a_{i,j}}\frac{K_i^{-2}-1}{q_i-q_i^{-1}}T^{-1}(-K_j^{-1}F_j)+[a_{i,j}]_{q_i}T^{-1}(-K_j^{-1}F_j)=$$

$$=\frac{q_i^{-a_{i,j}}-q_i^{a_{i,j}}}{q_i-q_i^{-1}}T^{-1}(-K_j^{-1}F_j)+[a_{i,j}]_{q_i}T^{-1}(-K_j^{-1}F_j)=0,$$

and the claim is stated. □

Proposition 11.
$\forall i\in I_0$

$$[E_i,\tilde{E}_{(m\delta,i)}]=-[2]_{q_i}\Bigg(\sum_{s=1}^{m-1}q_i^{-2(s-1)}(q_i-q_i^{-1})\tilde{E}_{((m-s)\delta,i)}T_{\omega_i^\vee}^{-s}(E_i)+$$

$$-q_i^{-2(m-1)}T_{\omega_i^\vee}^{-m}(E_i)\Bigg),$$

$$[\tilde{E}_{(m\delta,i)},-K_i^{-1}F_i]=$$

$$=-[2]_{q_i}\Bigg(\sum_{s=1}^{m-1}q_i^{-2(s-1)}(q_i-q_i^{-1})T_{\omega_i^\vee}^{s}(-K_i^{-1}F_i)\tilde{E}_{((m-s)\delta,i)}+$$

$$-q_i^{-2(m-1)}T_{\omega_i^\vee}^{m}(-K_i^{-1}F_i)\Bigg).$$

Proof: Recall that $\mathcal{U}_q(i)\cong\mathcal{U}_{q_i}(\widehat{sl(2)})$ and that under this isomorphism the images of $\tilde{E}_{(m\delta,i)}$, $T_{\omega_i^\vee}^{-r}(E_i)$ and $T_{\omega_i^\vee}^{r}(-K_i^{-1}F_i)$ (for $m>0$, $r\in\mathbb{Z}$) are respectively $\tilde{E}_{m\delta}$, $(T_0T_\tau)^{-r}(E_1)$ and $(T_0T_\tau)^{r}(-K_1^{-1}F_1)$ where τ is the nontrivial Dynkin diagram automorphism for $\widehat{sl(2)}$.
Then the thesis follows from example 1.C2.10. □

To sum up, the following commutation formulas hold:

Proposition 12.
$\forall i,j\in I_0$, $\forall m>0$, $\forall r\in\mathbb{Z}$

$$[T_{\omega_j^\vee}^{-r}(E_j),\tilde{E}_{(m\delta,i)}]=$$

$$=-[a_{i,j}]_{q_i}\Bigg(\sum_{s=1}^{m-1}(o(i)o(j))^s q_i^{-(s-1)a_{i,j}}(q_i-q_i^{-1})\tilde{E}_{((m-s)\delta,i)}T_{\omega_j^\vee}^{-r-s}(E_j)+$$

$$-(o(i)o(j))^m q_i^{-(m-1)a_{i,j}}T_{\omega_j^\vee}^{-r-m}(E_j)\Bigg)\quad\text{and}$$

$$[\tilde{E}_{(m\delta,i)},T_{\omega_j^\vee}^{r}(-K_j^{-1}F_j)]=$$

$$= -[a_{i,j}]_{q_i}\Big(\sum_{s=1}^{m-1}(o(i)o(j))^s q_i^{-(s-1)a_{i,j}}(q_i - q_i^{-1})T_{\omega_j^-}^{r+s}(-K_j^{-1}F_j)\tilde{E}_{((m-s)\delta,i)}+$$

$$-(o(i)o(j))^m q_i^{-(m-1)a_{i,j}}T_{\omega_j^-}^{r+m}(-K_j^{-1}F_j)\Big),$$

where $o : I_0 \to \{\pm 1\}$ is such that if $a_{i,j} < 0$ $(i, j \in I_0)$ then $o(i)o(j) = -1$.

Proof: I only need to remark that $o : I_0 \to \{\pm 1\}$ with the stated properties exists: this is obviously true and actually there exist exactly two such maps, o and $-o$. □

Lemma 13.

$$\sum_{m\geq 1}\frac{1}{m!}\Big(-\sum_{r>0}\frac{t^r - t^{-r}}{r}x^r\Big)^m = (1-t^2)\sum_{s>0}t^{-s}x^s,$$

or, equivalently,

$$\exp\Big(-\sum_{r>0}\frac{t^r - t^{-r}}{r}x^r\Big) = 1 + (1-t^2)\sum_{s>0}t^{-s}x^s.$$

Proof:

$$\exp\Big(-\sum_{r>0}\frac{t^r - t^{-r}}{r}x^r\Big) - 1 =$$

$$= \exp\Big(\sum_{r>0}\frac{(t^{-1}x)^r}{r} - \sum_{r>0}\frac{(tx)^r}{r}\Big) - 1 = \exp\big(\lg(1-tx) - \lg(1-t^{-1}x)\big) - 1 =$$

$$= \frac{1-tx}{1-t^{-1}x} - 1 = \frac{1 - tx - 1 + t^{-1}x}{1-t^{-1}x} =$$

$$= (t^{-1} - t)x\sum_{s\geq 0}(t^{-1}x)^s = (1-t^2)\sum_{s>0}t^{-s}x^s.$$

□

Lemma 14.
Let $\tilde{S}_m^{\pm}$ $(m > 0)$ and $T^{\pm 1}$ be commuting operators on a $\mathbb{C}(q)$-algebra A and let $v \in \mathbb{C}(q)^*$, $a \in \mathbb{Z}$ and $x \in A$ be such that

$$(\tilde{S}_m^+ - \tilde{S}_m^-)(x) =$$

$$= -[a]_v\Big(\sum_{s=1}^{m-1}v^{-(s-1)a}(v - v^{-1})T^s\tilde{S}_{m-s}^- - v^{-(m-1)a}T^m\Big)(x).$$

Define operators $S_m^{\pm}$ $(m > 0)$ by setting

$$S^{\pm}(u) \doteq \sum_{m>0} S_m^{\pm} u^m, \quad \tilde{S}^{\pm}(u) \doteq \sum_{m>0} \tilde{S}_m^{\pm} u^m$$

and requiring

$$1 - (v - v^{-1})\tilde{S}^{\pm}(u) = \exp(v - v^{-1})S^{\pm}(u).$$

Then $\forall m > 0$

$$(v - v^{-1})(S_m^{+} - S_m^{-})(x) = -(v - v^{-1})\frac{[ma]_v}{m}T^m(x),$$

so that, if $v \neq \pm 1$,

$$(S_m^{+} - S_m^{-})(x) = -\frac{[ma]_v}{m}T^m(x).$$

Proof: The hypotheses imply that

$$(v - v^{-1})(\tilde{S}^{+}(u) - \tilde{S}^{-}(u))(x) =$$

$$= (v^{-a} - v^{a})\sum_{m>0} u^m \left(\sum_{s=1}^{m-1} v^{-(s-1)a}(v - v^{-1})T^s \tilde{S}_{m-s}^{-} - v^{-(m-1)a}T^m\right)(x) =$$

$$= (1 - v^{2a})\sum_{s>0}\left(\sum_{m>s} u^m v^{-sa}(v - v^{-1})T^s \tilde{S}_{m-s}^{-} - v^{-sa}u^s T^s\right)(x) =$$

$$= (1 - v^{2a})\sum_{s>0}(uv^{-a}T)^s((v - v^{-1})\tilde{S}^{-}(u) - \mathrm{id})(x);$$

thus

$$(v - v^{-1})\tilde{S}^{+}(u)(x) = \Big(\mathrm{id} +$$

$$+\Big(\mathrm{id} + (1 - v^{2a})\sum_{s>0}(uv^{-a}T)^s\Big)((v - v^{-1})\tilde{S}^{-}(u) - \mathrm{id})\Big)(x) =$$

$$= \Big(\mathrm{id} + T'(u)((v - v^{-1})\tilde{S}^{-}(u) - \mathrm{id})\Big)(x)$$

where $T'(u) = \mathrm{id} + (1 - v^{2a})\sum_{s>0}(uv^{-a}T)^s$; in particular

$$(\mathrm{id} - (v - v^{-1})\tilde{S}^{+}(u))(x) = T'(u)(\mathrm{id} - (v - v^{-1})\tilde{S}^{-}(u))(x).$$

Since $\tilde{S}^{\pm}(u)T'(u) = T'(u)\tilde{S}^{\pm}(u)$ and $\tilde{S}^{+}(u)\tilde{S}^{-}(u) = \tilde{S}^{-}(u)\tilde{S}^{+}(u)$ one has that

$$\lg\big(\mathrm{id} - (v - v^{-1})\tilde{S}^{+}(u)\big)(x) = \lg\big(T'(u)(\mathrm{id} - (v - v^{-1})\tilde{S}^{-}(u))\big)(x).$$

But lemma 2.4.13 implies that

$$T'(u) = 1 + (1 - v^{2a}) \sum_{s>0} (uv^{-a}T)^s =$$

$$= \exp\left(- \sum_{m>0} \frac{v^{am} - v^{-am}}{m} (uT)^m \right).$$

Hence

$$(v - v^{-1})S^+(u)(x) = \lg\big(1 - (v - v^{-1})\tilde{S}^+(u)\big)(x) =$$

$$= \lg\big(T'(u)(1 - (v - v^{-1})\tilde{S}^-(u))\big)(x) =$$

$$= \lg(T'(u))(x) + \lg((1 - (v - v^{-1})\tilde{S}^-(u)))(x) =$$

$$= (v - v^{-1})\left(\dot{-} \sum_{m>0} \frac{[am]_v}{m} (uT)^m + S^-(u) \right)(x),$$

so that $\forall m > 0$

$$(v - v^{-1})(S_m^+ - S_m^-)(x) = -(v - v^{-1})\frac{[am]_v}{m}T^m(x);$$

in particular, if $v \neq \pm 1$, $(S_m^+ - S_m^-)(x) = -\frac{[am]_v}{m}T^m(x) \quad \forall m > 0.$ □

Proposition 15.
$\forall i, j \in I_0, \forall m > 0, \forall r \in \mathbb{Z}$

$$[T_{\omega_j^\vee}^{-r}(E_j), E_{(m\delta,i)}] = -(o(i)o(j))^m \frac{[ma_{i,j}]_{q_i}}{m} T_{\omega_j^\vee}^{-r-m}(E_j),$$

$$[E_{(m\delta,i)}, T_{\omega_j^\vee}^{r}(-K_j^{-1}F_j)] = -(o(i)o(j))^m \frac{[ma_{i,j}]_{q_i}}{m} T_{\omega_j^\vee}^{r+m}(-K_j^{-1}F_j).$$

Proof: $\forall m > 0$ let S_m^+ and $\tilde{S}_m^+$ be the operators on $\mathcal{U}_q$ of right multiplication by $E_{(m\delta,i)}$ and $\tilde{E}_{(m\delta,i)}$ respectively, and S_m^- and $\tilde{S}_m^-$ be the operators on $\mathcal{U}_q$ of left multiplication by $E_{(m\delta,i)}$ and $\tilde{E}_{(m\delta,i)}$ respectively. Then the definition and properties of $E_{(m\delta,i)}$ and $\tilde{E}_{(m\delta,i)}$ and proposition 2.4.12 imply that $\tilde{S}_m^\pm$ $(m > 0)$, $o(i)o(j)T_{\omega_j^\vee}^{\mp 1}$, q_i, $a_{i,j}$ and E_j verify the hypotheses of lemma 2.4.14, and so do $\tilde{S}_m^\mp$ $(m > 0)$, $o(i)o(j)T_{\omega_j^\vee}^{\pm 1}$, q_i, $a_{i,j}$ and $-K_j^{-1}F_j$. The claim then follows applying lemma 2.4.14.   □

I'm now ready to prove the commutation formulas $[E_{(r\delta,i)}, F_{(s\delta,j)}]$ for $r, s > 0$ and $i, j \in I_0$ such that $a_{i,j} \neq 0$.

Proposition 16.
$\forall i,j\in I_0\ \forall r\geq s>0$

$$[E_{(r\delta,i)},\tilde{F}_{(s\delta,j)}]=-\delta_{rs}(o(i)o(j))^r\frac{[ra_{i,j}]_{q_i}}{r}\frac{K_{r\delta}-K_{-r\delta}}{q_j-q_j^{-1}}.$$

Proof:

$$[E_{(r\delta,i)},\tilde{F}_{(s\delta,j)}]=[E_{(r\delta,i)},\Omega(\tilde{E}_{(s\delta,j)})]=$$

$$=[E_{(r\delta,i)},F_jT^s_{\omega_j^-}(E_jK_j)-q_j^2T^s_{\omega_j^-}(E_jK_j)F_j]=$$

$$=\big([E_{(r\delta,i)},F_j]T^s_{\omega_j^-}(E_j)+F_j[E_{(r\delta,i)},T^s_{\omega_j^-}(E_j)]+$$

$$-[E_{(r\delta,i)},T^s_{\omega_j^-}(E_j)]F_j-T^s_{\omega_j^-}(E_j)[E_{(r\delta,i)},F_j]\big)K_{\alpha_j-s\delta}=$$

$$=K_{\alpha_j-s\delta}\Bigg(-[K_j[E_{(r\delta,i)},-K_j^{-1}F_j],T^s_{\omega_j^-}(E_j)]+$$

$$+\frac{(o(i)o(j))^r[ra_{i,j}]_{q_i}}{r}[F_j,T^{s-r}_{\omega_j^-}(E_j)]\Bigg)=$$

$$=\frac{(o(i)o(j))^r[ra_{i,j}]_{q_i}}{r}K_{\alpha_j-s\delta}\big([K_jT^r_{\omega_j^-}(-K_j^{-1}F_j),T^s_{\omega_j^-}(E_j)]+K_j\tilde{E}_{((r-s)\delta,j)}\big)=$$

$$=\frac{(o(i)o(j))^r[ra_{i,j}]_{q_i}}{r}K_{2\alpha_j-s\delta}\big(T^s_{\omega_j^-}(T^{r-s}_{\omega_j^-}(-K_j^{-1}F_j)E_j+$$

$$-q_j^{-2}E_jT^{r-s}_{\omega_j^-}(-K_j^{-1}F_j))+\tilde{E}_{((r-s)\delta,j)}\big)=$$

$$=\frac{(o(i)o(j))^r[ra_{i,j}]_{q_i}}{r}K_{2\alpha_j-s\delta}(\tilde{E}_{((r-s)\delta,j)}-T^s_{\omega_j^-}(\tilde{E}_{((r-s)\delta,j)}))=$$

$$=-\delta_{rs}(o(i)o(j))^r\frac{[ra_{i,j}]_{q_i}}{r}\frac{K_{r\delta}-K_{-r\delta}}{q_j-q_j^{-1}}$$

because if $r\neq s$ $T^s_{\omega_j^-}(\tilde{E}_{((r-s)\delta,j)})=\tilde{E}_{((r-s)\delta,j)}$, while if $r=s$

$$K_{2\alpha_j-r\delta}(\tilde{E}_{(0,j)}-T^s_{\omega_j^-}(\tilde{E}_{(0,j)}))=$$

$$=K_{2\alpha_j-r\delta}\frac{K_j^{-2}-1-K^2_{r\delta-\alpha_j}+1}{q_j-q_j^{-1}}=-\frac{K_{r\delta}-K_{-r\delta}}{q_j-q_j^{-1}}.$$

□

Proposition 17.
$\forall i,j\in I_0\ \forall r,s>0$, $[E_{(r\delta,i)},F_{(s\delta,j)}]=\delta_{rs}(o(i)o(j))^r\frac{[ra_{i,j}]_{q_i}}{r}\frac{K_{r\delta}-K_{-r\delta}}{q_j-q_j^{-1}}$.

Proof: If $s < r$ the thesis is an immediate consequence of proposition 2.4.16 because $F_{(s\delta,j)}$ is a polynomial in $\tilde{F}_{(m\delta,j)}$ with $m \le s < r$.

$$\text{If} \quad s > r \quad [E_{(r\delta,i)}, F_{(s\delta,j)}] = \Omega[E_{(s\delta,j)}, F_{(r\delta,i)}] = 0.$$

Finally, if $s = r$ recall that

$$(q_j^{-1} - q_j)\sum_{k>0} F_{(k\delta,j)}u^k = \lg(1 - (q_j^{-1} - q_j)\sum_{k>0} \tilde{F}_{(k\delta,j)}u^k)$$

so that $(q_j^{-1} - q_j)F_{(r\delta,j)} = -(q_j^{-1} - q_j)\tilde{F}_{(r\delta,j)} + X$ where $[E_{(r\delta,i)}, X] = 0$ thanks to proposition 2.4.16; hence

$$[E_{(r\delta,i)}, F_{(r\delta,j)}] = -[E_{(r\delta,i)}, \tilde{F}_{(r\delta,j)}] = (o(i)o(j))^r \frac{[ra_{i,j}]_{q_i}}{r} \frac{K_{r\delta} - K_{-r\delta}}{q_j - q_j^{-1}}.$$

□

§5. Linear (and triangular) transformations of the imaginary root vectors.

Remark 1.
Corollary 2.3.4 implies that if there exist a good ordering $\prec$ of $\tilde{R}_+$ and elements $\bar{F}_\alpha \in \mathcal{U}^-_{q,-\alpha}$ $(\alpha \in \tilde{R}_+)$ satisfying the following properties:
1) $\forall \alpha \in R_+^{\rm re}$ $\bar{F}_\alpha \doteqdot F_\alpha$;
2) The monomials in $\{\bar{F}_{(m\delta,i)} | m > 0, i \in I_0\}$ form a $\mathbb{C}(q)$-basis of $\Omega(\mathcal{U}^+_{q,\rm im})$;
3) $\forall \alpha \in R_+^{\rm im} \times I_0$, $[E_\alpha, \bar{F}_\alpha] \in \mathcal{U}_q^0$;
4) $\forall \alpha, \beta \in R_+^{\rm im} \times I_0$ with $\alpha \succ \beta$ $(E_\alpha, \bar{F}_\beta)$ gives no contribution to the highest term;
then $\{E_\alpha | \alpha \in \tilde{R}_+\}$, $\{\bar{F}_\alpha | \alpha \in \tilde{R}_+\}$ and $\prec$ satisfy the conditions of proposition 2.2.6 while, $\forall \eta \in Q_+$,

$$\{E(\underline{\gamma}) | \underline{\gamma} \in \mathrm{Par}(\eta)\}, \quad \{F(\underline{\gamma}) | \underline{\gamma} \in \mathrm{Par}(\eta)\}, \quad \{\bar{F}(\underline{\gamma}) | \underline{\gamma} \in \mathrm{Par}(\eta)\} \quad \text{and} \quad \prec$$

satisfy the conditions of proposition 2.2.3.
Proof: Of course $\{\bar{F}(\underline{\gamma}) | \underline{\gamma} \in \cup_{\eta \in Q_+} \mathrm{Par}(\eta)\}$ is a basis of $\mathcal{U}_q^-$, hence $\forall \eta \in Q_+$ $\{\bar{F}(\underline{\gamma}) | \underline{\gamma} \in \mathrm{Par}(\eta)\}$ is a basis of $\mathcal{U}^-_{q,\eta}$, and the claim is proved. □

Remark 2.
The commutation formulas between $E_{(r\delta,i)}$ and $F_{(s\delta,j)}$ show that there is no ordering $\prec$ on $R_+^{\rm im} \times I_0$ such that $\forall \beta, \gamma \in R_+^{\rm im} \times I_0$ with $\beta \succ \gamma$, (E_β, F_γ) gives no contribution to the highest term.
Indeed, if $i, j \in I_0$ are such that $a_{i,j} \neq 0$, then $\forall m > 0$ both $(E_{(m\delta,i)}, F_{(m\delta,j)})$ and $(E_{(m\delta,j)}, F_{(m\delta,i)})$ give contribution to the highest term.

However the same formulas prove that

$$[E_{(r\delta,i)},x]\in\mathcal{U}_q^0 \ \ \forall x \ \text{ in the } \mathbb{C}(q)\text{−linear span of } \{F_{(r\delta,j)}|j\in I_0\}$$

$$\text{and } [E_{(r\delta,i)},x]=0 \ \ \forall x \ \text{ in the } \mathbb{C}(q)\text{−linear span of } \{F_{(s\delta,j)}|j\in I_0, s\neq r\}.$$

So the condition 2) of remark 2.5.1 can be replaced by the stronger following one:
2′) $\forall m>0$ $\{\bar F_{(m\delta,i)}|i\in I_0\}$ is a basis (over $\mathbb{C}(q)$) of the $\mathbb{C}(q)$-linear span of $\{F_{(m\delta,j)}|j\in I_0\}$.
Indeed in this case, $\forall m>0$, $\{F_{(m\delta,i)}|i\in I_0\}$ and $\{\bar F_{(m\delta,i)}|i\in I_0\}$ are bases of the same $\mathbb{C}(q)$-vector space, so that, since the monomials in

$$\{F_{(m\delta,i)}|m>0, i\in I_0\}$$

form a basis of $\mathcal{U}^+_{q,\mathrm{im}}$, also those in $\{\bar F_{(m\delta,i)}|m>0, i\in I_0\}$ do.
Notice that condition 3) of remark 2.5.1 is then automatically satisfied. Moreover $\forall r\neq s$ $\forall i,j\in I_0$ $[E_{(r\delta,i)},\bar F_{(s\delta,j)}]=0$. Thus the condition for property 4) of remark 2.5.1 to be satisfied depends only on the commutation rules, $\forall m>0$, between $E_{(m\delta,i)}$ and $\bar F_{(m\delta,j)}$ $(\forall i,j\in I_0)$.
Remark also that $[E_{(m\delta,i)},\bar F_{(m\delta,j)}]=a(K_{m\delta}-K_{-m\delta})$, so $(E_{(m\delta,i)},\bar F_{(m\delta,j)})$ gives no contribution to the highest term if and only if $a=0$, that is if and only if $[E_{(m\delta,i)},\bar F_{(m\delta,j)}]=0$. This means that if I call $\prec_m$ the ordering of I_0 given by $i\prec_m j \Leftrightarrow (m\delta,i)\prec(m\delta,j)$, this condition depends only on $\{\prec_m |m>0\}$, not on $\prec$.

Definition 3.
Let $\{\bar F_\alpha|\alpha\in\tilde R_+\}$ satisfy the conditions 1) of remark 2.5.1 and 2′) of remark 2.5.2. Define $A^{(m)}$ to be the matrix with coefficients in $\mathbb{C}(q)$ which represents the change of basis

$$\text{from } \{F_{(m\delta,i)}|i\in I_0\} \text{ to } \{\bar F_{(m\delta,i)}|i\in I_0\},$$

that is

$$\bar F_{(m\delta,i)}=\sum_{j\in I_0}A^{(m)}_{ij}F_{(m\delta,j)} \ \ \forall i\in I_0$$

Define $A(\eta)$ to be the matrix of change of basis of $\mathcal{U}^-_{q,-\eta}$

$$\text{from } \{F(\underline{\gamma})|\underline{\gamma}\in\mathrm{Par}(\eta)\} \text{ and } \{\bar F(\underline{\gamma})|\underline{\gamma}\in\mathrm{Par}(\eta)\},$$

that is

$$\bar F(\underline{\gamma})=\sum_{\underline{\tilde\gamma}\in\mathrm{Par}(\eta)}A(\eta)_{\underline{\gamma},\underline{\tilde\gamma}}F(\underline{\tilde\gamma}) \ \ \forall\underline{\gamma}\in\mathrm{Par}(\eta).$$

Remark 4.
If $\{\bar{F}_\alpha | \alpha \in \tilde{R}\}$ and $\prec$ satisfy the conditions 1) and 4) of remark 2.5.1 and 2′) of remark 2.5.2, proposition 2.2.3 implies that

$$b_\eta = \det A(\eta)^{-1} \det H_\eta^{\max}(\underline{E}, \underline{\bar{F}}),$$

where $H_\eta^{\max}(\underline{E}, \underline{\bar{F}})$ is diagonal; hence

$$b_\eta = \det A(\eta)^{-1} \prod_{\underline{\gamma} \in \mathrm{Par}(\eta)} H^{\max}(E(\underline{\gamma}), \bar{F}(\underline{\gamma})).$$

Then it is useful to find conditions on $A^{(m)}$ so that $\det A(\eta)$ can be easily calculated (for instance $A(\eta)$ is triangular). To this aim the following proposition holds.

Proposition 5.
If $A^{(m)}$ is upper triangular $\forall m > 0$, then $A(\eta)$ is upper triangular $\forall \eta \in Q_+$; moreover, $\forall \underline{\gamma} = (\gamma_1, ..., \gamma_r) \in \mathrm{Par}(\eta)$, $A(\eta)_{\underline{\gamma}, \underline{\gamma}} = \prod_{i=1}^r a_{\gamma_i}$ where

$$a_{\gamma_i} \doteq \begin{cases} 1 & \text{if } \gamma_i \in R_+^{\mathrm{re}} \\ A_{jj}^{(m)} & \text{if } \gamma_i = (m\delta, j). \end{cases}$$

In particular $\det A(\eta) = \prod_{\underline{\gamma} = (\gamma_1, ..., \gamma_r) \in \mathrm{Par}(\eta)} \prod_{i=1}^r a_{\gamma_i}$.

Proof: I define on $\mathrm{Par}(\eta)$ a relation $\mathcal{R}$ in the following way: if $\underline{\gamma} = (\gamma_1, ..., \gamma_r)$, $\underline{\tilde{\gamma}} = (\tilde{\gamma}_1, ..., \tilde{\gamma}_s)$, then

$$\underline{\gamma} \mathcal{R} \underline{\tilde{\gamma}} \Leftrightarrow \exists k \in \{1, ..., r\}, \tilde{k} \in \{1, ..., s\} \text{ such that } \gamma_k = (m\delta, i),\ \tilde{\gamma}_{\tilde{k}} = (m\delta, j)$$

with $A_{ij}^{(m)} \neq 0$, and $(\gamma_1, ..., \gamma_{k-1}, \gamma_{k+1}, ..., \gamma_r) = (\tilde{\gamma}_1, ..., \tilde{\gamma}_{\tilde{k}-1}, \tilde{\gamma}_{\tilde{k}+1}, ..., \tilde{\gamma}_s)$ (in particular $r = s$).
I will also write $\underline{\gamma} \mathcal{R}(\gamma_k, \tilde{\gamma}_{\tilde{k}}) \underline{\tilde{\gamma}}$ to stress the dependence on γ_k and $\tilde{\gamma}_{\tilde{k}}$.
Notice that $A(\eta)_{\underline{\gamma}, \underline{\tilde{\gamma}}} \neq 0$ if and only if $\#\{u | \gamma_u \notin R_+^{\mathrm{re}}\} = \#\{u | \tilde{\gamma}_u \notin R_+^{\mathrm{re}}\} = t$ and there exists a bijection $\sigma \in \mathfrak{S}_t$ such that, if

$$\{u_1, ..., u_t\} = \{u | \gamma_u \notin R_+^{\mathrm{re}}\} \text{ and } \{\tilde{u}_1, ..., \tilde{u}_t\} = \{u | \tilde{\gamma}_u \notin R_+^{\mathrm{re}}\},$$

$p(\tilde{\gamma}_{\tilde{u}_{\sigma(k)}}) = p(\gamma_{u_k})$ $\forall k = 1, ..., t$ and $\exists \underline{\gamma}^{(0)}, ..., \underline{\gamma}^{(t)} \in \mathrm{Par}(\eta)$ such that

$$\underline{\gamma}^{(0)} = \underline{\gamma},\ \underline{\gamma}^{(t)} = \underline{\tilde{\gamma}} \text{ and } \underline{\gamma}^{(k-1)} \mathcal{R}(\gamma_{u_k}, \tilde{\gamma}_{\tilde{u}_{\sigma(k)}}) \underline{\gamma}^{(k)} \quad \forall k = 1, ..., t.$$

Hence, in order to prove that $A(\eta)$ is upper triangular, it suffices to prove that $\underline{\gamma} \mathcal{R} \underline{\tilde{\gamma}} \Rightarrow \underline{\gamma} \preceq \underline{\tilde{\gamma}}$. But this is an immediate consequence of the fact that $A_{ij}^{(m)} \neq 0 \Rightarrow (m\delta, i) \preceq (m\delta, j)$.

To conclude notice that $\forall \underline{\gamma} \in \mathrm{Par}(\eta)$ $A(\eta)_{\underline{\gamma},\underline{\gamma}} \neq 0$ (because $A(\eta)$ is triangular and invertible); then

$$\underline{\gamma} = \underline{\gamma}^{(0)} \preceq \ldots \preceq \underline{\gamma}^{(t)} = \underline{\gamma}, \quad \text{so that} \quad \underline{\gamma}^{(k)} = \underline{\gamma} \quad \text{and} \quad \gamma_{u_k} = \tilde{\gamma}_{\tilde{u}_{\sigma(k)}} \quad \forall k = 1, \ldots, t.$$

Hence $A(\eta)_{\underline{\gamma},\underline{\gamma}} = \prod_{k=1}^{t} A_{i_k,i_k}^{(m_k)}$ where $\gamma_{u_k} = (m_k\delta, i_k)$. □

At this point I restrict my attention to looking for elements $\bar{F}_{(m\delta,i)}$ satisfying condition 4) of remark 2.5.1 and condition 2′) of remark 2.5.2, and such that the matrix $A^{(m)}$ is upper triangular.

Proposition 6.
The condition for the existence of elements $\bar{F}_\alpha$ satisfying conditions 4) of remark 2.5.1 and 2′) of remark 2.5.2, and the triangularity of $A^{(m)}$ $\forall m > 0$ is equivalent to the existence, $\forall m > 0$, of a solution of the system

$$\sum_{k \succeq_m j} A'^{(m)}_{j,k} [(\alpha_i|\alpha_k)]_{q^m} = 0 \quad \forall i, j \in I_0 \ \text{ s.t. } \ i \succ_m j, \quad \text{with} \quad A'^{(m)}_{j,j} \neq 0 \quad \forall j \in I_0,$$

the two being connected by

$$\bar{F}_{(m\delta,j)} = \sum_{k \succeq_m j} o(k)^m [d_k]_q A'^{(m)}_{j,k} F_{(m\delta,k)} \quad \forall m > 0, \ j \in I_0.$$

Proof: Let $A^{(m)} = \left(A^{(m)}_{i,j}\right)_{i,j \in I_0}$ be the matrix defined in definition 2.5.3.
The condition that $A^{(m)}$ be upper triangular is equivalent to saying that $\bar{F}_{(m\delta,j)} = \sum_{k \succeq_m j} A^{(m)}_{j,k} F_{(m\delta,k)}$ $\forall m > 0$, $j \in I_0$ and of course $A^{(m)}_{j,j} \neq 0$ $\forall m > 0$ $\forall j \in I_0$.
Moreover, thanks to remark 2.5.2, condition 4) of remark 2.5.1 holds if and only if $\forall m > 0$, $\forall i \succ_m j$ $[E_{(m\delta,i)}, \bar{F}_{(m\delta,j)}] = 0$; but

$$[E_{(m\delta,i)}, \bar{F}_{(m\delta,j)}] = \sum_{k \succeq_m j} A^{(m)}_{j,k} [E_{(m\delta,i)}, F_{(m\delta,k)}] =$$

$$= \frac{o(i)^m}{m} \sum_{k \succeq_m j} o(k)^m A^{(m)}_{j,k} [m a_{i,k}]_{q_i} \frac{K_{m\delta} - K_{-m\delta}}{q_k - q_k^{-1}} =$$

$$= \frac{o(i)^m [m]_q}{m [d_i]_q} \Big(\sum_{k \succeq_m j} A'^{(m)}_{j,k} [(\alpha_i|\alpha_k)]_{q^m} \Big) \frac{K_{m\delta} - K_{-m\delta}}{q - q^{-1}}$$

where $A'^{(m)}_{j,k} = \dfrac{o(k)^m A^{(m)}_{j,k}}{[d_k]_q}$. This immediately implies the claim because $A'^{(m)}_{j,k} = 0 \Leftrightarrow A^{(m)}_{j,k} = 0$. □

Remark 7.
I can consider a good ordering $\prec$ such that $\forall m > 0$ the ordering of I_0 $\prec_m$ is independent of m and is equal to the ordering $<$ induced by the bijection between I_0 and $\{1, ..., n\}$ given in the classification of the Dynkin diagrams, see §1.A4. (that is $(m\delta, r) \prec (m\delta, s) \Leftrightarrow r < s$ in $\mathbb{N}$). So I'm now reduced to look for a solution of the system

$$(*)\ \sum_{k \geq j} A'^{(m)}_{j,k}[(\alpha_i|\alpha_k)]_{q^m} = 0 \ \ \forall i, j \in I_0 \ \text{ s.t. } \ i > j, \quad \text{with} \quad A'^{(m)}_{j,j} \neq 0 \ \ \forall j \in I_0.$$

§6. Solution of the triangular system for new imaginary root vectors.

From now on $\{A'^{(m)}_{j,k}\}$ will denote a solution of $(*)$.

Proposition 1.
If $(i, j) \in I_0^2$ is such that $i \geq j$ and $\forall r > i$ the Dynkin diagram restricted to $\{k \in I_0 | k \geq j$ and $(\alpha_r|\alpha_k) \neq 0\}$ is either $\underset{r-1}{\circ}\!-\!\underset{r}{\circ}\!-\!\underset{r+1}{\circ}$ if $r < n$ or $\underset{n-1}{\circ}\!-\!\underset{n}{\circ}$ if $r = n$, then $A'^{(m)}_{j,i} = [n - i + 1]_{q^{md_i}} A'^{(m)}_{j,n}$.

Proof: First notice that the fact that (i, j) satisfies the conditions of the proposition means that the Dynkin diagram restricted to the vertices $s \geq i$ is $\underset{i}{\circ}\!-\!\underset{i+1}{\circ}\cdots\underset{n-1}{\circ}\!-\!\underset{n}{\circ}$ and $\forall r, s \in I_0$ with $r > i > s \geq j$ the vertices r and s are not connected, so that (r, j) also satisfies the conditions of the proposition $\forall r \geq i$ $(r \in I_0)$. I use induction on $n - i$:
i) if $i = n$ the thesis is obvious;
ii) let $i = n - 1$; then

$$0 = \sum_{k \geq j} A'^{(m)}_{j,k}[(\alpha_n|\alpha_k)]_{q^m} = -A'^{(m)}_{j,n-1}[d_n]_{q^m} + A'^{(m)}_{j,n}[2d_n]_{q^m}$$

and the thesis follows noticing that $d_n = d_{n-1} = d_i$ and $\dfrac{[2d_i]_{q^m}}{[d_i]_{q^m}} = [2]_{q^{md_i}}$;
iii) let $i < n - 1$; then

$$0 = \sum_{k \geq j} A'^{(m)}_{j,k}[(\alpha_{i+1}|\alpha_k)]_{q^m} =$$

$$= -A'^{(m)}_{j,i}[d_{i+1}]_{q^m} + A'^{(m)}_{j,i+1}[2d_{i+1}]_{q^m} - A'^{(m)}_{j,i+2}[d_{i+1}]_{q^m} =$$

$$= [d_i]_{q^m}(-A'^{(m)}_{j,i} + [2]_{q^{md_i}} A'^{(m)}_{j,i+1} - A'^{(m)}_{j,i+2}) \ \text{ because } \ d_{i+1} = d_i.$$

Hence

$$A'^{(m)}_{j,i} = [2]_{q^{md_i}} A'^{(m)}_{j,i+1} - A'^{(m)}_{j,i+2} =$$

$$= ([2]_{q^{md_i}}[n - i]_{q^{md_i}} - [n - i - 1]_{q^{md_i}})A'^{(m)}_{j,n} = [n - i + 1]_{q^{md_i}} A'^{(m)}_{j,n}. \qquad \square$$

Proposition 2.
If $(i,j)\in I_0^2$ are such that $i\geq j$, $(i+1,j)$ satisfies the condition of proposition 2.6.1 and (i,j) does not, and if moreover $\{k\geq j|a_{k,i+1}\neq 0\}=\{i,i+1,i+2\}$, then $[-a_{i+1,i}]_{q^{md_{i+1}}}A'^{(m)}_{j,i}=[n-i+1]_{q^{md_{i+1}}}A'^{(m)}_{j,n}$.

Proof: Notice that the situation described in the hypothesis is that the Dynkin diagram restricted to the vertices $s\geq i$ is one of the following:

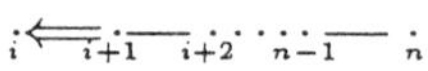 with $i=1$ or 2,

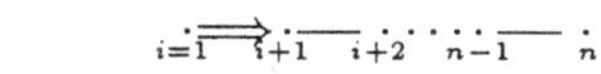

or

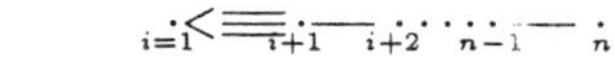

and, when $i=2$, $a_{1,r}=0\ \forall r>2$.
Then
$$0=\sum_{k\geq j}A'^{(m)}_{j,k}[(\alpha_{i+1}|\alpha_k)]_{q^m}=$$
$$=A'^{(m)}_{j,i}[(\alpha_{i+1}|\alpha_i)]_{q^m}+A'^{(m)}_{j,i+1}[(\alpha_{i+1}|\alpha_{i+1})]_{q^m}+A'^{(m)}_{j,i+2}[(\alpha_{i+1}|\alpha_{i+2})]_{q^m}=$$
$$=[d_{i+1}]_{q^m}(A'^{(m)}_{j,i}[a_{i+1,i}]_{q^{md_{i+1}}}+A'^{(m)}_{j,i+1}[2]_{q^{md_{i+1}}}-A'^{(m)}_{j,i+2}.$$
Hence
$$[-a_{i+1,i}]_{q^{md_{i+1}}}A'^{(m)}_{j,i}=$$
$$=([2]_{q^{md_{i+1}}}[n-i]_{q^{md_{i+1}}}-[n-i-1]_{q^{md_{i+1}}})A'^{(m)}_{j,n}=[n-i+1]_{q^{md_{i+1}}}A'^{(m)}_{j,n},$$
which is the thesis. □

Proposition 3.
If $i\in I_0$ is such that $\{k\in I_0|a_{i+1,k}\neq 0\}=\{1,i,i+1,i+2\}$ then $\forall j\leq i$
$$[i]_{q^m}A'^{(m)}_{1,j}=\begin{cases}[j-1]_{q^m}[n-i]_{q^m}A'^{(m)}_{1,n} & \text{if } 1<j\leq i\\ [n]_{q^m}A'^{(m)}_{1,n} & \text{if } j=1.\end{cases}$$

Proof: Remark that this situation happens only if $\mathfrak{g}=D_n$, $i=2$ or $\mathfrak{g}=E_n$, $i=3$ and that in these cases if $j>1$ or $r>i$ (and $j\leq r$), (r,j) satisfies the conditions required in proposition 2.6.1, in particular $(i+1,i)$ satisfies the conditions of proposition 2.6.1.
Then if $i\in I_0$ is as stated, $d_j=1\ \forall j$ and $(\alpha_j|\alpha_k)\in\{0,-1\}\ \forall j,k\in I_0$. So
$$0=\sum_{k\in I_0}A'^{(m)}_{1,k}[(\alpha_{i+1}|\alpha_k)]_{q^m}=-A'^{(m)}_{1,1}-A'^{(m)}_{1,i}+A'^{(m)}_{1,i+1}[2]_{q^m}-A'^{(m)}_{1,i+2},$$

hence $A'^{(m)}_{1,1} + A'^{(m)}_{1,i} = A'^{(m)}_{1,i+1}[2]_{q^m} - A'^{(m)}_{1,i+2} =$

$$= ([2]_{q^m}[n-i]_{q^m} - [n-i-1]_{q^m})A'^{(m)}_{1,n} = [n-i+1]_{q^m}A'^{(m)}_{1,n}.$$

On the other hand if $1 < j \leq i$

$$\{k \in I_0 | a_{k,j} \neq 0\} = \{k \in I_0 | (\alpha_k|\alpha_j) \in \{-1, 2\}\} = \{j-1, j, j+1\} \cap (I_0 \setminus \{1\}),$$

so that

$$0 = \sum_{k \in I_0} A'^{(m)}_{1,k}[(\alpha_j|\alpha_k)]_{q^m} =$$

$$= (1-\delta_{j2})A'^{(m)}_{1,j-1}[(\alpha_j|\alpha_{j-1})]_{q^m} + A'^{(m)}_{1,j}[(\alpha_j|\alpha_j)]_{q^m} + A'^{(m)}_{1,j+1}[(\alpha_j|\alpha_{j+1})]_{q^m} =$$

$$= -(1-\delta_{j2})A'^{(m)}_{1,j-1} + [2]_{q^m}A'^{(m)}_{1,j} - A'^{(m)}_{1,j+1}.$$

This is equivalent to saying that

$$\begin{cases} [2]_{q^m}A'^{(m)}_{1,2} = A'^{(m)}_{1,3} = [n-2]_{q^m}A'^{(m)}_{1,n} \\ A'^{(m)}_{1,1} + A'^{(m)}_{1,2} = [n-1]_{q^m}A'^{(m)}_{1,n} \end{cases} \quad \text{if } i = 2$$

$$\begin{cases} [2]_{q^m}A'^{(m)}_{1,2} = A'^{(m)}_{1,3} \\ A'^{(m)}_{1,1} + A'^{(m)}_{1,3} = [n-2]_{q^m}A'^{(m)}_{1,n} \\ A'^{(m)}_{1,2} = [2]_{q^m}A'^{(m)}_{1,3} - A'^{(m)}_{1,4} = [2]_{q^m}A'^{(m)}_{1,3} - [n-3]_{q^m}A'^{(m)}_{1,n} \end{cases} \quad \text{if } i = 3.$$

Thus, if $i = 2$,

$$\begin{cases} [2]_{q^m}A'^{(m)}_{1,2} = [n-2]_{q^m}A'^{(m)}_{1,n} \\ [2]_{q^m}A'^{(m)}_{1,1} = [n]_{q^m}A'^{(m)}_{1,n} \end{cases}$$

while, if $i = 3$,

$$\begin{cases} [3]_{q^m}A'^{(m)}_{1,3} = [2]_{q^m}[n-3]_{q^m}A'^{(m)}_{1,n} \\ [3]_{q^m}A'^{(m)}_{1,2} = [n-3]_{q^m}A'^{(m)}_{1,n} \\ [3]_{q^m}A'^{(m)}_{1,1} = ([3]_{q^m}[n-2]_{q^m} - [2]_{q^m}[n-3]_{q^m})A'^{(m)}_{1,n} = [n]_{q^m}A'^{(m)}_{1,n} \end{cases}$$

and this is the thesis. □

Proposition 4.
Suppose $\mathfrak{g} = F_4$; then $A'^{(m)}_{1,1} = (q^{5m} + q^{-5m})A'^{(m)}_{1,4}$.
Proof:

$$0 = \sum_{k \geq 1} A'^{(m)}_{1,k}[(\alpha_2|\alpha_k)]_{q^m} =$$

$$= A'^{(m)}_{1,1}[(\alpha_2|\alpha_1)]_{q^m} + A'^{(m)}_{1,2}[(\alpha_2|\alpha_2)]_{q^m} + A'^{(m)}_{1,3}[(\alpha_2|\alpha_3)]_{q^m} =$$

$$= -A'^{(m)}_{1,1} + [2]_{q^m} A'^{(m)}_{1,2} - [2]_{q^m} A'^{(m)}_{1,3};$$

hence

$$A'^{(m)}_{1,1} = [2]_{q^m}(A'^{(m)}_{1,2} - A'^{(m)}_{1,3}) = [2]_{q^m}([3]_{q^{2m}} - [2]_{q^{2m}})A'^{(m)}_{1,4} =$$

$$= ([2\cdot 3]_{q^m} - [2\cdot 2]_{q^m})A'^{(m)}_{1,4} =$$

$$= \frac{q^{6m} - q^{-6m} - q^{4m} + q^{-4m}}{q^m - q^{-m}} A'^{(m)}_{1,4} = (q^{5m} + q^{-5m})A'^{(m)}_{1,4}.$$

□

I shall now collect propositions from 2.6.1 to 2.6.4 to get a solution of $(*)$.

Proposition 5.
$\forall\{b_i^{(m)}|i\in I_0, m>0\}\subset \mathbb{C}(q)^*$ the set $\{A'^{(m)}_{i,j}|i,j\in I_0, j\geq i, m>0\}$ given by

$$A'^{(m)}_{j,k} = \begin{cases} [n-k+1]_{q^{md_k}} b_j^{(m)} & \text{if } (k,j) \text{ is as in proposition 2.6.1} \\ \dfrac{[n-k+1]_{q^{md_{k+1}}}}{[-a_{k+1,k}]_{q^{md_{k+1}}}} b_j^{(m)} & \text{if } (k,j) \text{ is as in proposition 2.6.2} \\ \dfrac{[k-1]_{q^m}[n-r]_{q^m}}{[r]_{q^m}} b_1^{(m)} & \text{if } j=1<k\leq r \text{ and } \mathfrak{g}=D_n\ (r=2) \text{ or } \mathfrak{g}=E_n\ (r=3). \\ \dfrac{[n]_{q^m}}{[r]_{q^m}} b_1^{(m)} & \text{if } j=k=1, \text{and } \mathfrak{g}=D_n\ (r=2) \text{ or } \mathfrak{g}=E_n\ (r=3) \\ [2]_{q^{5m}} b_1^{(m)} & \text{if } j=k=1,\ \mathfrak{g}=F_4 \end{cases}$$

is a solution of the system $(*)$, and all the solutions of $(*)$ are of this form.

Proof: Remark that the set of solution of $(*)$ is in one-to-one correspondence with the product of the sets of solutions of

$$(*)_j^{(m)} \qquad \sum_{k\geq j} A'^{(m)}_{j,k}[(\alpha_i|\alpha_k)]_{q^m} = 0 \text{ for } i>j, \quad A'^{(m)}_{j,j}\neq 0$$

because the set of unknowns appearing in the system $(*)_j^{(m)}$ is disjoint from the set of unknowns appearing in the system $(*)_{\tilde{j}}^{(\tilde{m})}$ $\forall(\tilde{m},\tilde{j})\neq(m,j)$.

Notice that propositions from 2.6.1 to 2.6.4 imply that $\forall j \in I_0, m > 0$, for each $b_j^{(m)} \in \mathbb{C}(q)$ there is at most one solution

$$\{A'^{(m)}_{j,k} | k \in I_0\} \text{ of } (*) \text{ such that } A^{(m)}_{j,n} = b_j^{(m)},$$

and this must be the one given above.
Moreover, since $A^{(m)}_{j,j}$ is a multiple of $A^{(m)}_{j,n}$, one necessarily has $A^{(m)}_{j,n} \neq 0$ for all the solutions of $(*)_j^{(m)}$.
Hence the unicity assertion is proved.
For the existence note that $\forall j \in I_0\ \forall m > 0$,

$$(\tilde{*})_j^{(m)} \qquad \sum_{k \geq j} A'^{(m)}_{j,k} [(\alpha_i|\alpha_k)]_{q^m} = 0 \quad \text{for } i > j$$

is a system of $n-j$ homogeneous equations in $n-j+1$ unknowns; hence it has a nontrivial solution (remark that all the solutions of $(*)_j^{(m)}$ are nontrivial solutions of $(\tilde{*})_j^{(m)}$).
But $A^{(m)}_{j,k}$ is a multiple of $A^{(m)}_{j,n}$ $\forall k \geq j$, hence the nontrivial solutions are those where $A^{(m)}_{j,n} \neq 0$, and are then indicized by $\mathbb{C}(q)^*$, setting

$$A^{(m)}_{j,n} = b_j^{(m)} \in \mathbb{C}(q).$$

Such a solution is a solution of $(*)_j^{(m)}$ because $A^{(m)}_{j,j}$ is a non zero multiple of $A^{(m)}_{j,n}$, hence different from zero.
The claim is then completely proved. □

Definition 6.
Given $\{b_j^{(m)} | j \in I_0, m > 0\} \subset \mathbb{C}(q)^*$ and the related (see proposition 2.6.5) solution of $(*)$ $\{A^{(m)}_{j,k} | j, k \in I_0, k \geq j, m > 0\}$, the elements $\bar{c}_i^{(m)}$ are defined, $\forall i \in I_0, m > 0$, by

$$\bar{c}_i^{(m)} \doteqdot \sum_{j \geq i} A'^{(m)}_{i,j} [(\alpha_i|\alpha_j)]_{q^m}.$$

Proposition 7.
Given $\{b_j^{(m)} | j \in I_0, m > 0\} \subset \mathbb{C}(q)^*$ and the related (see proposition 2.6.5) solution of $(*)$

$$\{A^{(m)}_{j,k} | j, k \in I_0, k \geq j, m > 0\},$$

the elements $\bar{c}_i^{(m)}$ with $i > 1$, $m > 0$ are given by

$$\bar{c}_i^{(m)} = \begin{cases} A'^{(m)}_{i-1,i-1} \dfrac{[d_i]_{q^m} b_i^{(m)}}{b_{i-1}^{(m)}} & \text{if } (i-1,i-1) \text{ is as in proposition} \\ & \text{2.6.1 or } i = 2, \mathfrak{g} = F_4 \\ A'^{(m)}_{i-1,i-1} \dfrac{[-d_i a_{i,i-1}]_{q^m} b_i^{(m)}}{b_{i-1}^{(m)}} & \text{if } (i-1,i-1) \text{ is as in} \\ & \text{proposition 2.6.2} \\ A'^{(m)}_{1,1} \dfrac{[r]_{q^m} b_2^{(m)}}{b_1^{(m)}} & \text{if } i = 2, \ \mathfrak{g} = D_n \text{ or } E_n, \\ & \text{and } r \text{ is as in proposition 2.6.3;} \end{cases}$$

the elements $\bar{c}_1^{(m)}$ for $m > 0$ are given in the following table:

$A_n^{(1)}$	$\bar{c}_1^{(m)} = [n+1]_{q^m} b_1^{(m)}$
$B_n^{(1)}$	$\bar{c}_1^{(m)} = [2]_{q^{m(2n-1)}} b_1^{(m)}$
$C_n^{(1)}$	$\bar{c}_1^{(m)} = [2]_{q^{m(n+1)}} b_1^{(m)}$
$D_n^{(1)}$	$\bar{c}_1^{(m)} = [2]_{q^{m(n-1)}} b_1^{(m)}$
$E_n^{(1)}$	$\bar{c}_1^{(m)} = \dfrac{[2]_{q^m}[2]_{q^{m(n-1)}} - [n-5]_{q^m}}{[3]_{q^m}} b_1^{(m)}$
$F_4^{(1)}$	$\bar{c}_1^{(m)} = \dfrac{[2]_{q^{9m}}}{[2]_{q^{3m}}} b_1^{(m)}$
$G_2^{(1)}$	$\bar{c}_1^{(m)} = \dfrac{[2]_{q^{6m}}}{[2]_{q^{2m}}} b_1^{(m)}.$

Proof: Let (i,i) satisfy the conditions of proposition 2.6.1 so that

$$A'^{(m)}_{i,k} = [n-k+1]_{q^{md_k}} b_i^{(m)} \quad \forall k \geq i \quad \text{and} \quad d_i = d_{i+1}.$$

$$\text{Then} \quad \bar{c}_i^{(m)} = \sum_{j \geq i} A'^{(m)}_{i,j} [(\alpha_i|\alpha_j)]_{q^m} =$$

$$= b_i^{(m)} [d_i]_{q^m} \left([n-i+1]_{q^{md_i}} [2]_{q^{m d_i}} - [n-i]_{q^{md_i}} \right) =$$

$$= b_i^{(m)} [d_i]_{q^m} [n-i+2]_{q^{md_i}}.$$

If $i > 1$ and $(i-1,i-1)$ satisfies the conditions of proposition 2.6.1, $d_i = d_{i-1}$ and $A'^{(m)}_{i-1,i-1} = [n-i+2]_{q^{md_{i-1}}} b_{i-1}^{(m)}$, hence

$$\bar{c}_i^{(m)} = A'^{(m)}_{i-1,i-1} \frac{[d_i]_{q^m} b_i^{(m)}}{b_{i-1}^{(m)}}.$$

If $i > 1$ and $(i-1, i-1)$ satisfies the conditions of proposition 2.6.2,

$$A'^{(m)}_{i-1,i-1} = \frac{[n-i+2]_{q^{md_i}}}{[-a_{i,i-1}]_{q^{md_i}}} b^{(m)}_{i-1},$$

so that

$$\bar{c}^{(m)}_i = A'^{(m)}_{i-1,i-1} \frac{[-(\alpha_i|\alpha_{i-1})]_{q^m} b^{(m)}_i}{b^{(m)}_{i-1}}.$$

If $\mathfrak{g} = D_n$ or E_n then $d_2 = 1$ and $A'^{(m)}_{1,1} = \frac{[n]_{q^m}}{[r]_{q^m}} b^{(m)}_1$ where r satisfies the hypotheses of proposition 2.6.3, from which $\bar{c}^{(m)}_2 = A'^{(m)}_{1,1} \frac{[r]_{q^m} b^{(m)}_2}{b^{(m)}_1}$.

Suppose now $\mathfrak{g} = F_4$; then

$$\bar{c}^{(m)}_2 = A'^{(m)}_{2,2}[(\alpha_2|\alpha_2)]_{q^m} + A'^{(m)}_{2,3}[(\alpha_2|\alpha_3)]_{q^m} =$$

$$= b^{(m)}_2 [2]_{q^m}([3]_{q^{2m}} - [2]_{q^{2m}}) = b^{(m)}_2 [2]_{q^{5m}} = A'^{(m)}_{1,1} \frac{b^{(m)}_2}{b^{(m)}_1}.$$

Finally if $\hat{\mathfrak{g}} = A^{(1)}_n$ $\bar{c}^{(m)}_1 = [n+1]_{q^m} b^{(m)}_1$.

Let us now study $\bar{c}^{(m)}_1$ when $\mathfrak{g} \neq A_n$.

If $\mathfrak{g} = B_n, C_n, G_2$,

$$\bar{c}^{(m)}_1 = A^{(m)}_{1,1}[2d_1]_{q^m} + A^{(m)}_{1,2}[(\alpha_1|\alpha_2)]_{q^m} =$$

$$= \frac{[d_1]_{q^m} b^{(m)}_1}{[-a_{2,1}]_{q^{md_2}}} \left([n]_{q^{md_2}}[2]_{q^{md_1}} - [n-1]_{q^{md_3}}[-a_{1,2}]_{q^{md_1}}[-a_{2,1}]_{q^{md_2}}\right) =$$

$$= b^{(m)}_1 ([n]_{q^{md_2}}[2]_{q^{md_1}} - [n-1]_{q^{md_2}}[\max(d_1, d_2)]_{q^m}) =$$

$$= \begin{cases} b^{(m)}_1([2n]_{q^m} - [2(n-1)]_{q^m}) = [2]_{q^{m(2n-1)}} b^{(m)}_1 & \text{if } \mathfrak{g} = B_n \\ b^{(m)}_1([n]_{q^m}[2]_{q^{2m}} - [n-1]_{q^m}[2]_{q^m}) = [2]_{q^{m(n+1)}} b^{(m)}_1 & \text{if } \mathfrak{g} = C_n \\ b^{(m)}_1([2]_{q^{3m}}[2]_{q^m} - [3]_{q^m}) = \frac{[2]_{q^{6m}}}{[2]_{q^{2m}}} b^{(m)}_1 & \text{if } \mathfrak{g} = G_2; \end{cases}$$

if $\mathfrak{g} = D_n$ or E_n and r satisfies the conditions of proposition 2.6.3,

$$\bar{c}^{(m)}_1 = A'^{(m)}_{1,1}[2]_{q^m} - A'^{(m)}_{1,r+1} =$$

$$= b^{(m)}_1 ([r]_{q^m})^{-1}\left([n]_{q^m}[2]_{q^m} - [r]_{q^m}[n-r]_{q^m}\right) =$$

$$= \begin{cases} b_1^{(m)}([n]_{q^m} - [n-2]_{q^m}) = [2]_{q^{m(n-1)}} b_1^{(m)} & \text{if } \mathfrak{g} = D_n \\ b_1^{(m)}([3]_{q^m})^{-1}\big([n]_{q^m}[2]_{q^m} - [3]_{q^m}[n-3]_{q^m}\big) = & \\ \quad = \dfrac{q^{mn} + q^{m(n-2)} - [n-5]_{q^m} + q^{-m(n-2)} + q^{-mn}}{[3]_{q^m}} b_1^{(m)} & \text{if } \mathfrak{g} = E_n; \end{cases}$$

finally, if $\mathfrak{g} = F_4$, we have that $\bar{c}_1^{(m)} = A_{1,1}'^{(m)}[2]_{q^m} - A_{1,2}'^{(m)} =$

$$= b_1^{(m)}(q^{6m} + q^{4m} + q^{-4m} + q^{-6m} - q^{4m} - 1 - q^{-4m}) = \frac{[2]_{q^{9m}}}{[2]_{q^{3m}}}.$$

□

Proposition 8.
There is exactly one choice of $\{b_j^{(m)} | j \in I_0, m > 0\} \subset \mathbb{C}(q)^*$ such that

$$A_{n,n}^{(m)} = 1 \quad \forall m > 0 \quad \text{and} \quad A_{i,i}^{(m)} = \bar{c}_{i+1}^{(m)} = \quad \forall i \in I_0, i < n.$$

Proof: Thanks to proposition 2.6.7, the condition that

$$A_{i,i}^{(m)} = \bar{c}_{i+1}^{(m)} \quad \forall i \in I_0, i < n$$

is equivalent to solving, $\forall m > 0$, the following linear system for the $b_i^{(m)}$'s:

$$b_i^{(m)} = \begin{cases} [d_{i+1}]_{q^m} b_{i+1}^{(m)} & \text{if } i < n \text{ and for } (i,i) \\ & \text{proposition 2.6.1 holds, or } i = 1, \mathfrak{g} = F_4 \\ [-d_{i+1}a_{i+1,i}]_{q^m} b_{i+1}^{(m)} & \text{if } i < n \text{ and for } (i,i) \\ & \text{proposition 2.6.2 holds} \\ [r]_{q^m} b_2^{(m)} & \text{if } i = 1 \text{ and } \mathfrak{g} = D_n, E_n \\ & (r \text{ as in proposition 2.6.3}) \end{cases}$$

(with $b_i^{(m)} \neq 0 \ \forall i \in I_0 \ \forall m > 0$).
Such a system has non trivial solutions because it is a homogeneous system of $n-1$ equations in n unknowns, and a solution is necessarily such that $\forall i < n$

$$b_i^{(m)} = \begin{cases} [d_n]_{q^m}^{n-i} b_n^{(m)} & \text{if for } (i,i) \text{ proposition 2.6.1 holds} \\ [-d_{i+1}a_{i+1,i}]_{q^m}[d_n]_{q^m}^{n-i-1} b_n^{(m)} & \text{if for } (i,i) \text{ proposition 2.6.2} \\ & \text{holds, or } i = 1, \mathfrak{g} = F_4 \\ [r]_{q^m} b_n^{(m)} & \text{if } i = 1 \text{ and } \mathfrak{g} = D_n, E_n \\ & (r \text{ as in proposition 2.6.3}). \end{cases}$$

Hence a solution depends uniquely on $b_n^{(m)}$, and $b_i^{(m)} \neq 0$ $\forall i \in I_0$ if and only if $b_n^{(m)} \neq 0$. This implies that $\forall \tilde{b}^{(m)} \in \mathbb{C}(q)^*$ there is exactly one solution such that $b_n^{(m)} = \tilde{b}^{(m)}$ (and all the solutions are of this form).

On the other hand, $\forall m > 0$ the condition that $A_{n,n}^{(m)} = 1$ is equivalent to saying $b_n^{(m)} = 1$, and the claim follows. □

From now on $\{A_{i,j}^{\prime(m)}\}$ will indicate the solution of the system $(*)$ such that $A_{n,n}^{\prime(m)} = 1$ $\forall m > 0$ and $A_{i,i}^{(m)} = c_{i+1}^{(m)}$ $\forall m > 0$, $\forall i \in I_0$ with $i < n$; this solution is resumed in the following table:

$$A_n^{(1)} \qquad A_{j,k}^{\prime(m)} = [n-k+1]_{q^m} \quad \forall k \geq j$$
$$\bar{c}_1^{(m)} = [n+1]_{q^m}$$

$$B_n^{(1)} \qquad A_{j,k}^{\prime(m)} = [n-k+1]_{q^{2m}}[2]_{q^m}^{n-j} \quad \forall k \geq j$$
$$\bar{c}_1^{(m)} = [2]_{q^m}^{n-1}[2]_{q^{m(2n-1)}}$$

$$C_n^{(1)} \qquad A_{j,k}^{\prime(m)} = \begin{cases} [n-k+1]_{q^m} & \text{if } k \geq j > 1 \text{ or } k = j = 1 \\ [n-k+1]_{q^m}[2]_{q^m} & \text{if } k > j = 1 \end{cases}$$
$$\bar{c}_1^{(m)} = [2]_{q^m}[2]_{q^{m(n+1)}}$$

$$D_n^{(1)} \qquad A_{j,k}^{\prime(m)} = \begin{cases} [n-k+1]_{q^m} & \text{if } k \geq j > 1 \text{ or } k = j = 1 \\ [n-k+1]_{q^m}[2]_{q^m} & \text{if } k > 2 = j+1 \\ [n-2]_{q^m} & \text{if } k = 2,\ j = 1 \end{cases}$$
$$\bar{c}_1^{(m)} = [2]_{q^m}[2]_{q^{m(n-1)}}$$

$$E_n^{(1)} \qquad A_{j,k}^{\prime(m)} = \begin{cases} [n-k+1]_{q^m} & \text{if } k \geq j > 1 \text{ or } k = j = 1 \\ [k-1]_{q^m}[n-3]_{q^m} & \text{if } j = 1 < k < 4 \\ [n-k+1]_{q^m}[3]_{q^m} & \text{if } j = 1,\ k \geq 4 \end{cases}$$
$$\bar{c}_1^{(m)} = \begin{cases} [9-n]_{q^m}[2]_{q^{3(n-4)m}}([2]_{q^{(n-4)m}})^{-1} & \text{if } n = 6,7 \\ [2]_{q^m}[2]_{q^{7m}} - [3]_{q^m} & \text{if } n = 8 \end{cases}$$

$$F_4^{(1)} \qquad A_{j,k}^{\prime(m)} = \begin{cases} [5-k]_{q^{2m}}[2]_{q^m}^{4-j} & \text{if } k \geq j > 1 \\ [2]_{q^m}^2[2]_{q^{5m}} & \text{if } k = j = 1 \\ [5-k]_{q^{2m}}[2]_{q^m}^2 & \text{if } k > j = 1 \end{cases}$$
$$\bar{c}_1^{(m)} = \frac{[2]_{q^m}^2[2]_{q^{9m}}}{[2]_{q^{3m}}}$$

$$G_2^{(1)} \qquad A_{j,k}^{\prime(m)} = \begin{cases} [3-k]_{q^{3m}}[3]_{q^m} & \text{if } k \geq j = 1 \\ 1 & \text{if } k = j = 2 \end{cases}$$
$$\bar{c}_1^{(m)} = \frac{[3]_{q^m}[2]_{q^{6m}}}{[2]_{q^{2m}}}.$$

Remark 9.
Notice that $\forall m > 0$ $\forall i, j \in I_0$ with $i \geq j$, both $A_{j,i}^{(m)}$ and $\bar{c}_i^{(m)}$ (and in particular $\bar{c}_1^{(m)}$) are in $\mathbb{C}[q, q^{-1}]$. □

Definition 10.
$\forall \alpha \in \tilde{R}_+$ define

$$\bar{F_\alpha} \doteqdot \begin{cases} F_\alpha & \text{if } \alpha \in R_+^{\text{re}} \\ \sum_{j \geq i} o(j)^m [d_j]_q A_{i,j}^{\prime(m)} F_{(m\delta,j)} & \text{if } \alpha = (m\delta, i). \end{cases}$$

§7. The highest coefficient b_η: complete description.

Theorem 1.
If $\prec$ is a good ordering of $\tilde{R}_+$ such that $\forall m > 0$ $(m\delta, i) \prec (m\delta, j) \Rightarrow i < j$ in I_0, the sets $\{E_\alpha | \alpha \in \tilde{R}_+\}$ and $\{\bar{F}_\alpha | \alpha \in \tilde{R}_+\}$ satisfy the conditions of proposition 2.2.6; in particular detH_η is the product of $\det(H_\eta(\underline{E}, \underline{\bar{F}}))$ times the inverse of $\prod_{\underline{\gamma}=(\gamma_1,\dots,\gamma_r)\in \mathrm{Par}(\eta)} \prod_{i=1}^r a_{\gamma_i}$, where

$$a_\alpha \doteqdot \begin{cases} 1 & \text{if } \alpha \in R_+^{\text{re}} \\ o(j)^m [d_j]_q A_{j,j}^{\prime(m)} & \text{if } \alpha = (m\delta, j). \end{cases}$$

Proof: Propositions 2.2.6, 2.4.17 and 2.5.6 imply that $\forall \underline{\gamma}, \underline{\tilde{\gamma}} \in \mathrm{Par}(\eta)$ such that $\underline{\gamma} \succ \underline{\tilde{\gamma}}$, $(E_{\underline{\gamma}}, \bar{F}_{\underline{\tilde{\gamma}}})$ gives no contribution to the highest term.
Then it is enough to apply proposition 2.5.5. □

Proposition 2.
$\forall \alpha \in \tilde{R}_+$

$$[E_\alpha, \bar{F}_\alpha] = \begin{cases} \dfrac{K_\alpha - K_\alpha^{-1}}{q_\alpha - q_\alpha^{-1}} & \text{if } \alpha \in R_+^{\text{re}} \\ \dfrac{o(i)^m [m]_q}{m} \bar{c}_i^{(m)} \dfrac{K_{p(\alpha)} - K_{p(\alpha)}^{-1}}{q_\alpha - q_\alpha^{-1}} & \text{if } \alpha = (m\delta, i) \in R_+^{\text{im}} \times I_0 \end{cases}$$

where $q_{(m\delta,i)} \doteqdot q_i$.
Proof: If $\alpha \in R_+^{\text{re}}$ see proposition 2.3.1.
If $\alpha = (m\delta, i)$ then

$$[E_\alpha, \bar{F}_\alpha] = [E_{(m\delta,i)}, \sum_{j \geq i} o(j)^m [d_j]_q A_{i,j}^{\prime(m)} F_{(m\delta,j)}] =$$

$$= \sum_{j \geq i} o(i)^m [d_j]_q A_{i,j}^{\prime(m)} \frac{[m(\alpha_i|\alpha_j)]_q}{m[d_i]_q} \frac{K_{m\delta} - K_{-m\delta}}{q_j - q_j^{-1}} =$$

$$= \frac{o(i)^m[m]_q}{m}\Big(\sum_{j\geq i} A'^{(m)}_{i,j}[(\alpha_i|\alpha_j)]_{q^m}\Big)\frac{K_{m\delta}-K_{-m\delta}}{q_i-q_i^{-1}} =$$

$$= \frac{o(i)^m[m]_q}{m}\bar{c}_i^{(m)}\frac{K_{m\delta}-K_{-m\delta}}{q_{(m\delta,i)}-q_{(m\delta,i)}^{-1}}.$$

□

Definition 3.
$\forall \underline{\gamma}=(\gamma_1,...,\gamma_r)\in \mathrm{Par}(\eta)$ $\forall\alpha\in\tilde{R}_+$, I denote with $\mu^\alpha_{\underline{\gamma}}$ the multiplicity of α in $\underline{\gamma}$, that is

$$\mu^\alpha_{\underline{\gamma}} \doteq \#\{i\in\{1,...,r\}|\gamma_i=\alpha\}.$$

Proposition 4.
$\forall\eta\in Q_+$, $\forall\underline{\gamma}=(\gamma_1,...,\gamma_r)\in\mathrm{Par}(\eta)$ the highest coefficient of $\pi([E_{\underline{\gamma}},\bar{F}(\underline{\gamma})])$ is, up to units of $\mathbb{C}[q,q^{-1}]$,

$$\prod_{\alpha\in\tilde{R}_+}\Big(\frac{c_\alpha}{q_\alpha-q_\alpha^{-1}}\Big)^{\mu^\alpha_{\underline{\gamma}}}\cdot\prod_{\alpha\in R^{\mathrm{re}}_+}[\mu^\alpha_{\underline{\gamma}}]_{q_\alpha}!$$

where $c_\alpha \doteq \begin{cases} 1 & \text{if } \alpha\in R^{\mathrm{re}}_+ \\ \frac{o(i)^m[m]_q}{m}\bar{c}_i^{(m)} & \text{if } \alpha=(m\delta,i). \end{cases}$

Proof: Induction on $\mathrm{ht}(\eta)$.
If $\mathrm{ht}(\eta)$=0 the thesis is obvious; let $\mathrm{ht}(\eta)>0$; then, if $\underline{\gamma}^*=(\gamma_2,..,\gamma_r)$,

$$\pi([E(\underline{\gamma}),\bar{F}(\underline{\gamma})])=\pi([E(\underline{\gamma}),\bar{F}_{\gamma_1}\bar{F}(\underline{\gamma}^*)])=$$

$$=\pi([E(\underline{\gamma}),\bar{F}_{\gamma_1}]\bar{F}(\underline{\gamma}^*))=$$

$$=\sum_{i=1}^r \pi(E_{\gamma_1}\cdot\ldots\cdot E_{\gamma_{i-1}}[E_{\gamma_i},\bar{F}_{\gamma_1}]E_{\gamma_{i+1}}\cdot\ldots\cdot E_{\gamma_r}\bar{F}(\underline{\gamma}^*))=$$

$$=\sum_{\substack{i=1\\ \gamma_i\preceq\gamma_1}}^r \pi(E_{\gamma_1}\cdot\ldots\cdot E_{\gamma_{i-1}}[E_{\gamma_i},\bar{F}_{\gamma_1}]E_{\gamma_{i+1}}\cdot\ldots\cdot E_{\gamma_r}\bar{F}(\underline{\gamma}^*))+\text{lower terms}=$$

$$=\sum_{\substack{i=1\\ \gamma_i=\gamma_1}}^r c_{\gamma_1}\pi\Big(E_{\gamma_1}\cdot\ldots\cdot E_{\gamma_{i-1}}\frac{K_{p(\gamma_1)}-K^{-1}_{p(\gamma_1)}}{q_{\gamma_1}-q_{\gamma_1}^{-1}}E_{\gamma_{i+1}}\cdot\ldots\cdot E_{\gamma_r}\bar{F}(\underline{\gamma}^*)\Big)+$$

$$+\text{lower terms}=$$

$$=\sum_{\substack{i=1\\ \gamma_i=\gamma_1}}^r \frac{c_{\gamma_1}q^{-(i-1)(p(\gamma_1)|p(\gamma_1))}}{q_{\gamma_1}-q_{\gamma_1}^{-1}}K_{p(\gamma_1)}\pi(E(\underline{\gamma}^*)\bar{F}(\underline{\gamma}^*))+\text{lower terms}=$$

$$= \frac{c_{\gamma_1}}{q_{\gamma_1} - q_{\gamma_1}^{-1}} \sum_{\substack{i=1\\ \gamma_i=\gamma_1}}^{r} q^{-(i-1)(p(\gamma_1)|p(\gamma_1))} K_{p(\gamma_1)} \pi([E(\underline{\gamma}^*), \bar{F}(\underline{\gamma}^*)]).$$

Now, if $\gamma_1 \in R_+^{\rm re}$,

$$\sum_{\substack{i=1\\ \gamma_i=\gamma_1}}^{r} q^{-(i-1)(\gamma_1|\gamma_1)} = 1 + q^{-(\gamma_1|\gamma_1)} + \ldots + q^{-(\mu_{\underline{\gamma}}^{\gamma_1}-1)(\gamma_1|\gamma_1)} =$$

$$= q^{-(\mu_{\underline{\gamma}}^{\gamma_1}-1)d_{\gamma_1}} [\mu_{\underline{\gamma}}^{\gamma_1}]_{q_{\gamma_1}},$$

while, if $\gamma_1 \in R_+^{\rm im} \times I_0$, $(p(\gamma_1)|p(\gamma_1)) = 0\ \forall i$; hence

$$\pi([E(\underline{\gamma}), \bar{F}(\underline{\gamma})]) = u_{\gamma_1} \frac{c_{\gamma_1}}{q_{\gamma_1} - q_{\gamma_1}^{-1}} K_{p(\gamma_1)} \pi([E(\underline{\gamma}^*), \bar{F}(\underline{\gamma}^*)]) + \text{lower terms}$$

where u_{γ_1} is, up to a unit in $\mathbb{C}[q, q^{-1}]$, $[\mu_{\underline{\gamma}}^{\gamma_1}]_{q_{\gamma_1}}$ if $\gamma_1 \in R_+^{\rm re}$ and 1 if $\gamma_1 \notin R_+^{\rm re}$. Since $\underline{\gamma}^* \in {\rm Par}(\eta - p(\gamma_1))$ and $\eta - p(\gamma_1) < \eta$ the inductive hypothesis applies, and the thesis then follows remarking that $\mu_{\underline{\gamma}}^{\beta} = \mu_{\underline{\gamma}^*}^{\beta} + \delta_{\beta,\gamma_1}$. □

Theorem 5.
The coefficient b_η of $\det H_\eta$ is, up to invertible elements of $\mathbb{C}[q, q^{-1}]$,

$$b_\eta = \prod_{\substack{\alpha \in \tilde{R}_+\\ m>0}} \left(\frac{[m]_{q_\alpha}}{q_\alpha - q_\alpha^{-1}}\right)^{{\rm par}(\eta - mp(\alpha))} \prod_{r,m>0} \left(\frac{\bar{c}_1^{(m)}}{\prod_{i \in I_0} [d_i]_{q^m}}\right)^{{\rm par}(\eta - mr\delta)}.$$

Proof: The arguments developed till now show that if a_α and c_α are as defined respectively in lemma 2.5.5 (or in theorem 2.7.1) and in proposition 2.7.4,

$$b_\eta = \prod_{\underline{\gamma} \in {\rm Par}(\eta)} \left(\prod_{\alpha \in \tilde{R}_+} \left(\frac{c_\alpha}{a_\alpha(q_\alpha - q_\alpha^{-1})}\right)^{\mu_{\underline{\gamma}}^{\alpha}} \prod_{\alpha \in R_+^{\rm re}} [\mu_{\underline{\gamma}}^{\alpha}]_{q_\alpha}! \right) =$$

$$= \left(\prod_{\alpha \in \tilde{R}_+} \left(\frac{c_\alpha}{a_\alpha(q_\alpha - q_\alpha^{-1})}\right)^{\sum_{\underline{\gamma} \in {\rm Par}(\eta)} \mu_{\underline{\gamma}}^{\alpha}} \right) \left(\prod_{\substack{\alpha \in R_+^{\rm re}\\ m>0}} ([m]_{q_\alpha}!)^{\#\{\underline{\gamma} \in {\rm Par}(\eta) | \mu_{\underline{\gamma}}^{\alpha} = m\}} \right) =$$

$$= \prod_{\substack{\alpha \in \tilde{R}_+\\ m>0}} \left(\frac{c_\alpha}{a_\alpha(q_\alpha - q_\alpha^{-1})}\right)^{m\#\{\underline{\gamma} \in {\rm Par}(\eta) | \mu_{\underline{\gamma}}^{\alpha} = m\}} \prod_{\substack{\alpha \in R_+^{\rm re}\\ m>0}} ([m]_{q_\alpha})^{\#\{\underline{\gamma} \in {\rm Par}(\eta) | \mu_{\underline{\gamma}}^{\alpha} \geq m\}}.$$

I shall now concentrate in calculating, $\forall \alpha \in \tilde{R}_+$, $\#\{\underline{\gamma} \in {\rm Par}(\eta) | \mu_{\underline{\gamma}}^{\alpha} = m\}$.

$\forall\eta\in Q_+$ $\forall\alpha\in\tilde{R}_+$ define a map ϑ_α from $\mathrm{Par}(\eta-p(\alpha))$ into $\mathrm{Par}(\eta)$ as follows:

$$\vartheta_\alpha(\gamma_1,...,\gamma_r)\doteq(\gamma_1,...,\gamma_{i-1},\alpha,\gamma_i,...,\gamma_r)$$

where $\gamma_{i-1}\preceq\alpha\preceq\gamma_i$.

Notice that ϑ_α is well defined and injective and that, $\forall\alpha,\beta\in\tilde{R}_+$, $\forall\underline{\gamma}\in\mathrm{Par}(\eta)$, $\mu^\alpha_{\vartheta_\beta(\underline{\gamma})}=\delta_{\beta,\alpha}+\mu^\alpha_{\underline{\gamma}}$; moreover if $\mu^\alpha_{\underline{\gamma}}\geq 1$ then $\underline{\gamma}\in\mathrm{Im}(\vartheta_\alpha)$.

More precisely if $\underline{\gamma}\in\mathrm{Par}(\eta)$, $\alpha\in\tilde{R}_+$ and $m\in\mathbb{N}$ then $\mu^\alpha_{\underline{\gamma}}=m\Leftrightarrow\underline{\gamma}\in\mathrm{Im}(\vartheta^m_\alpha)$ and $\underline{\gamma}\notin\mathrm{Im}(\vartheta^{m+1}_\alpha)$, so that $\forall\alpha\in\tilde{R}_+$, $\forall m\in\mathbb{N}$

$$\#\{\underline{\gamma}\in\mathrm{Par}(\eta)|\mu^\alpha_{\underline{\gamma}}=m\}=\mathrm{par}(\eta-mp(\alpha))-\mathrm{par}(\eta-(m+1)p(\alpha))$$

and $\forall\alpha\in\tilde{R}_+$, $\forall m\in\mathbb{N}$, $\#\{\underline{\gamma}\in\mathrm{Par}(\eta)|\mu^\alpha_{\underline{\gamma}}\geq m\}=\mathrm{par}(\eta-mp(\alpha))$. Hence

$$b_\eta=\prod_{\substack{\alpha\in\tilde{R}_+\\ m>0}}\left(\frac{c_\alpha}{a_\alpha(q_\alpha-q_\alpha^{-1})}\right)^{m(\mathrm{par}(\eta-mp(\alpha))-\mathrm{par}(\eta-(m+1)p(\alpha)))}\cdot$$

$$\cdot\prod_{\substack{\alpha\in R^{\mathrm{re}}_+\\ m>0}}[m]_{q_\alpha}^{\mathrm{par}(\eta-m\alpha)}=$$

$$=\prod_{\substack{\alpha\in\tilde{R}_+\\ m>0}}\left(\frac{c_\alpha}{a_\alpha(q_\alpha-q_\alpha^{-1})}\right)^{\mathrm{par}(\eta-mp(\alpha))}\prod_{\substack{\alpha\in R^{\mathrm{re}}_+\\ m>0}}[m]_{q_\alpha}^{\mathrm{par}(\eta-m\alpha)}=$$

$$=\prod_{\substack{\alpha\in R^{\mathrm{re}}_+\\ m>0}}\left(\frac{[m]_{q_\alpha}}{q_\alpha-q_\alpha^{-1}}\right)^{\mathrm{par}(\eta-m\alpha)}\prod_{\substack{r,m>0\\ i\in I_0}}\left(\frac{c_{(r\delta,i)}}{a_{(r\delta,i)}(q_i-q_i^{-1})}\right)^{\mathrm{par}(\eta-mr\delta)}$$

because $a_\alpha=c_\alpha=1$ if $\alpha\in R^{\mathrm{re}}_+$.

Moreover $\forall r>0$

$$\prod_{i\in I_0}\frac{c_{(r\delta,i)}}{a_{(r\delta,i)}}=u\prod_{i\in I_0}\frac{[r]_q\bar{c}_i^{(r)}}{[d_i]_qA'^{(r)}_{i,i}}=$$

$$=u\left(\prod_{i\in I_0}\frac{[r]_q}{[d_i]_q}\right)\frac{\bar{c}_1^{(r)}}{A'^{(r)}_{n,n}}\prod_{i=1}^{n-1}\frac{\bar{c}_{i+1}^{(r)}}{A'^{(r)}_{i,i}}=u\bar{c}_1^{(r)}\prod_{i\in I_0}\frac{[r]_q}{[d_i]_q}=u\bar{c}_1^{(r)}\prod_{i\in I_0}\frac{[r]_{q_i}}{[d_i]_{q^r}}$$

where $u\in\mathbb{Q}^*$; hence

$$b_\eta=\prod_{\substack{\alpha\in R^{\mathrm{re}}_+\\ m>0}}\left(\frac{[m]_{q_\alpha}}{q_\alpha-q_\alpha^{-1}}\right)^{\mathrm{par}(\eta-m\alpha)}\cdot$$

$$\cdot\prod_{r,m>0}\left(\bar{c}_1^{(r)}\prod_{i\in I_0}\frac{[r]_{q_i}}{[d_i]_{q^r}(q_i-q_i^{-1})}\right)^{\mathrm{par}(\eta-mr\delta)}=$$

$$=\prod_{\substack{\alpha\in\tilde{R}_+\\ m>0}}\left(\frac{[m]_{q_\alpha}}{q_\alpha-q_\alpha^{-1}}\right)^{\mathrm{par}(\eta-mp(\alpha))}\prod_{r,m>0}\left(\frac{\bar{c}_1^{(m)}}{\prod_{i\in I_0}[d_i]_{q^m}}\right)^{\mathrm{par}(\eta-mr\delta)}.$$

□

CHAPTER 3. THE CENTER $\mathcal{Z}(\mathcal{U}_\varepsilon)$ AT ODD ROOTS OF 1.

In this chapter I want to describe the center of $\mathcal{U}_\varepsilon$, where $\varepsilon\in\mathbb{C}^*$ is a primitive l^{th} root of 1, with l odd number bigger than 1 (bigger than 3 if $\mathfrak{g}=G_2$).
To this aim I'll show that there is a strict connection between $\mathcal{Z}(\mathcal{U}_\varepsilon)\cap\mathcal{U}^+_{\varepsilon,\eta}$ and the multiplicity of ε in $\det H_\eta$; more precisely the multiplicity of ε in $\det H_\eta$ determines an upper bound for the dimension of $\mathcal{Z}(\mathcal{U}_\varepsilon)\cap\mathcal{U}^+_{\varepsilon,\eta}$, so that knowing the multiplicity of ε in $\det H_\eta$ $\forall\eta\in Q_+$ and exhibiting explicitely some elements of $\mathcal{Z}(\mathcal{U}_\varepsilon)\cap\mathcal{U}^+_\varepsilon$ give enough information for a complete description of $\mathcal{Z}(\mathcal{U}_\varepsilon)\cap\mathcal{U}^+_\varepsilon$.
But proposition 2.1.10 and theorem 2.7.5 will be shown to be enough to find the exact multiplicity of ε in $\det H_\eta$ $\forall\varepsilon\in\mathbb{C}^*$, so that the problem is to find the right number of central vectors.
The commutation rules between the root vectors allow to find some central "root" vectors. When $\mathfrak{g}=A_n$ or E_6 some degeneracies occur for certain values of l, which however will be also described, using the imaginary root vectors $\bar{F}_\alpha$.
The final result of this thesis consists in describing $\mathcal{Z}(\mathcal{U}_\varepsilon)$ as the tensor product of the algebras $\mathcal{Z}(\mathcal{U}_\varepsilon)\cap\mathcal{U}^-_\varepsilon$, $\mathcal{Z}(\mathcal{U}_\varepsilon)\cap\mathcal{U}^0_\varepsilon$ and $\mathcal{Z}(\mathcal{U}_\varepsilon)\cap\mathcal{U}^+_\varepsilon$, that is, knowing the positive, negative and null part of the center is enough to describe it completely: thus $\mathcal{Z}(\mathcal{U}_\varepsilon)$ will be given explicitely.

§1. The multiplicity of ε in $\det H_\eta$.

Definition 1.
Let $P\in\mathcal{A}[K_0^{\pm1},...,K_n^{\pm1}]$ be any non zero polynomial, and let ε be any non zero complex number such that $\varepsilon^r\neq\varepsilon^{-r}$ $\forall r\in\{1,...,\max(d_0,...,d_n)\}$.
The multiplicity of ε in P is defined to be the maximal $m\in\mathbb{N}$ such that $(q-\varepsilon)^m|P$ in $\mathcal{A}[K_0^{\pm1},...,K_n^{\pm1}]$.

Remark 2.
1) The multiplicity of ε in P is well defined since $q-\varepsilon$ is not invertible in $\mathcal{A}$;
2) The multiplicity of ε in P is the least between the multiplicities of its coefficients;
3) The multiplicity of ε in PQ is the sum of the multiplicities of ε in P and in Q.

Proof: 1) follows from the fact that $q-\varepsilon$ is not invertible in $\mathcal{A}$; otherwise there would exist $r\leq\max(d_0,...,d_n)$ $(r>0)$ such that $\varepsilon^r=\varepsilon^{-r}$, which would contradict the choice of ε.
2) Let m and m' be respectively the multiplicity of ε in P and the least between the multiplicities of the coefficients of P; of course $(q-\varepsilon)^{m'}$ divides all the coefficients of P, hence it divides P and $m'\leq m$; on the other hand $(q-\varepsilon)^{-m}P$ has coefficients in $\mathcal{A}$, hence $(q-\varepsilon)^m$ divides all the coefficients of P, that is $m\leq m'$; thus $m=m'$, which is the claim.
3) immediately follows from the irreducibility of $q-\varepsilon$ in $\mathcal{A}$. □

Proposition 3.
Let ε be as in definition 3.1.1. Then $\forall \eta \in Q$ the multiplicity of ε in $\det H_\eta$ is equal to the multiplicity of ε in b_η.

Proof: $\frac{\det H_\eta}{b_\eta}$ is a polynomial with coefficients in $\mathbb{C}[q,q^{-1}]$ (see proposition 2.1.10) whose leading coefficient is 1 (by definition of b_η, see definition 2.1.8). Then, thanks to remark 3.1.2 2), the multiplicity of ε in $\frac{\det H_\eta}{b_\eta}$ is zero. This implies the claim, thanks to remark 3.1.2 3). □

Remark 4.
In this section l will denote an odd number bigger than $\max\{d_0, ..., d_n\}$ (which is equal to $\max\{a_{i,j} | i,j \in I_0\}$), that is $l > 3$ if $\mathfrak{g} = G_2$ and $l > 1$ otherwise; ε will denote a primitive l^{th} root of 1.
Of course ε is such that $\varepsilon^r \neq \varepsilon^{-r}$ $\forall r \in \{1, ..., \max(d_0, ..., d_n)\}$.

Definition 5.
$\forall \alpha \in \tilde{R}_+$ define $d_\alpha \doteq d_i$ if $\exists w \in W$ such that $\alpha = w(\alpha_i)$ or if $\exists m > 0$ such that $\alpha = (m\delta, i)$; notice that $q_\alpha = q^{d_\alpha}$.
Also denote by l_α the integer $l_\alpha \doteq \frac{l}{\text{g.c.d.}(d_\alpha, l)}$ and $\forall i \in I_0$ $l_i \doteq \frac{l}{\text{g.c.d.}(d_i, l)}$.

Lemma 6.
Let $\alpha \in \tilde{R}_+$, $m, r > 0$; then
1) $(q-\varepsilon) \nmid (q_\alpha - q_\alpha^{-1})$;
2) $(q-\varepsilon) | [m]_{q_\alpha} \Leftrightarrow l | 2md_\alpha \Leftrightarrow l_\alpha | m$;
3) $(q-\varepsilon)^2 \nmid [m]_{q^r}$;
4) $(q-\varepsilon) | [d_i]_{q^m} \Leftrightarrow l | 2md_i, l \nmid 2m \Leftrightarrow d_i = 3$, $3 | l$ and $m = \frac{lk}{3}$ with $3 \nmid k$.

Proof: $(q-\varepsilon) | (q^r - q^{-r}) \Leftrightarrow l | 2r$; moreover $(q-\varepsilon)^2$ never divides $(q^r - q^{-r})$ (unless $r = 0$); the claim is then immediate. □

Before going on I recall here the table of the d_i's:

$$A_n^{(1)}: \quad d_i = 1 \ \forall i \in I_0$$

$$B_n^{(1)}: \quad d_i = \begin{cases} 1 & \text{if } i = 1 \\ 2 & \text{otherwise} \end{cases}$$

$$C_n^{(1)}: \quad d_i = \begin{cases} 2 & \text{if } i = 1 \\ 1 & \text{otherwise} \end{cases}$$

$$D_n^{(1)}: \quad d_i = 1 \ \forall i \in I_0$$

$$E_n^{(1)}: \quad d_i = 1 \ \forall i \in I_0$$

$$F_4^{(1)}: \quad d_1 = d_2 = 1, \quad d_3 = d_4 = 2$$

$$G_2^{(1)}: \quad d_1 = 1, \quad d_2 = 3.$$

Corollary 7.
Let $N_{\varepsilon,\eta}$ be the multiplicity of ε in $\det H_\eta$. Then

$$N_{\varepsilon,\eta} = \sum_{\substack{\alpha\in\hat{R}_+\\ m>0}} \mathrm{par}(\eta - ml_\alpha p(\alpha)) - \sum_{\substack{r,m>0:\\ 3\nmid m\\ i:d_i=3\\ l'\in\mathbb{Z}:l=3l'}} \mathrm{par}(\eta - ml_i r\delta) + C_{\varepsilon,\eta}$$

where $C_{\varepsilon,\eta}$ is the multiplicity of ε in

$$\prod_{r,m>0} \left(\bar{c}_1^{(m)}\right)^{\mathrm{par}(\eta-mr\delta)}.$$

Proof: The proof follows immediately from lemma 3.1.6, proposition 3.1.3 and theorem 2.7.5. □

Proposition 8.
In the notations of corollary 3.1.7,

$$C_{\varepsilon,\eta} = \begin{cases} \sum_{\substack{r,m>0:\\ \text{g.c.d.}(l,n+1)\nmid m}} \mathrm{par}\Big(\eta - \frac{mlr}{\text{g.c.d.}(l,n+1)}\delta\Big) & \text{if } \mathfrak{g} = A_n \\ \sum_{\substack{r,m>0:\\ 3\nmid m}} \mathrm{par}\big(\eta - \frac{mlr}{3}\delta\big) & \text{if } \mathfrak{g} = E_6 \text{ or} \\ & \mathfrak{g} = G_2 \text{ and } 3|l \\ 0 & \text{otherwise.} \end{cases}$$

Proof: I notice that $(q-\varepsilon) \nmid [2]_{q^k}$ $\forall k > 0$; hence $C_{\varepsilon,\eta} = 0$ if $\mathfrak{g} \neq A_n, E_6, E_8, G_2$.
Moreover $(q-\varepsilon) \nmid \bar{c}_1^{(m)}$ $\forall m > 0$ if $\mathfrak{g} = E_8$, so that also in this case $C_{\varepsilon,\eta} = 0$.
Now, if $\mathfrak{g} = A_n$, $(q-\varepsilon)^2 \nmid \bar{c}_1^{(m)}$ $\forall m > 0$ and

$$(q-\varepsilon)|\bar{c}_1^{(m)} \Leftrightarrow (q-\varepsilon)|[n+1]_{q^m} \Leftrightarrow \frac{l}{\text{g.c.d.}(l,n+1)}|m, l\nmid m.$$

Hence, if $\mathfrak{g} = A_n$,

$$C_{\varepsilon,\eta} = \sum_{\substack{r,m>0:\\ \text{g.c.d.}(l,n+1)\nmid m}} \mathrm{par}\Big(\eta - \frac{mlr}{\text{g.c.d.}(l,n+1)}\delta\Big).$$

Finally, if $\mathfrak{g} = E_6$ or G_2, $(q-\varepsilon)^2 \nmid \bar{c}_1^{(m)}$ $\forall m > 0$ and

$$(q-\varepsilon)|\bar{c}_1^{(m)} \Leftrightarrow (q-\varepsilon)|[3]_{q^m} \Leftrightarrow \frac{l}{\text{g.c.d.}(3,l)}|m, l\nmid m;$$

thus, if $\mathfrak{g} = E_6$ or G_2,

$$C_{\varepsilon,\eta} = \begin{cases} 0 & \text{if } 3 \nmid l \\ \sum_{\substack{r,m>0:\\ 3\nmid m}} \operatorname{par}\left(\eta - \frac{mlr}{3}\delta\right) & \text{if } 3|l. \end{cases}$$

□

Theorem 9.

$$N_{\varepsilon,\eta} = \sum_{\substack{\alpha\in\tilde{R}_+\\ m>0}} \operatorname{par}(\eta - ml_\alpha p(\alpha)) + \tilde{C}_{\varepsilon,\eta}$$

where $\tilde{C}_{\varepsilon,\eta}$ is given by:

$$\tilde{C}_{\varepsilon,\eta} = \begin{cases} \sum_{\substack{r,m>0:\\ \text{g.c.d.}(l,n+1)\nmid m}} \operatorname{par}\left(\eta - \frac{mlr}{\text{g.c.d.}(l,n+1)}\delta\right) & \text{if } \mathfrak{g} = A_n \\ \sum_{\substack{r,m>0:\\ 3\nmid m}} \operatorname{par}\left(\eta - \frac{mlr}{3}\delta\right) & \text{if } \mathfrak{g} = E_6 \text{ and } 3|l \\ 0 & \text{otherwise.} \end{cases}$$

Proof: Since $\exists i\in I_0$ such that $d_i = 3$ if and only if $\mathfrak{g} = G_2$, the thesis follows remarking that in this case

$$\sum_{\substack{r,m>0:\\ 3\nmid m}} \operatorname{par}\left(\eta - \frac{mlr}{3}\delta\right) = \sum_{\substack{r,m>0:\\ 3\nmid m\\ i\in I_0:d_i=3}} \operatorname{par}\left(\eta - \frac{mlr}{3}\delta\right).$$

□

Remark 10.

I notice that, since l is odd, $l_\alpha = \begin{cases} \frac{l}{3} & \text{if } \mathfrak{g} = G_2,\ d_\alpha = 3|l \\ l & \text{otherwise.} \end{cases}$

§2. Connections between the multiplicity of ε in $\det H_\eta$ and $\mathcal{Z}(\mathcal{U}_\varepsilon) \cap \mathcal{U}^+_{\varepsilon,\eta}$.

In the preceding section I found the highest power of $(q-\varepsilon)$ dividing $\det H_\eta$. In this section I want to study the relations between $\det H_\eta$ and $\mathcal{Z}(\mathcal{U}_\varepsilon)$. More exactly, I give conditions to get an upper bound for $\dim_{\mathbb{C}}(\mathcal{Z}(\mathcal{U}_\varepsilon) \cap \mathcal{U}^+_{\varepsilon,\eta})$.
I start this section proving a general lemma which I'll need in the following.

Lemma 1.
Let $\mathcal{V}$ be a $\mathbb{C}$-algebra with no zero divisors and suppose that there exist subsets S', S'' of $\mathcal{V}$, a linear ordering $\prec$ of S', subsets $J', T' \subseteq S'$ and $J'', T'' \subseteq S''$, and a map $f' : J' \to \mathbb{Z}_+$, such that:
i) $\{s'_1 \cdot \ldots \cdot s'_r s'' | s'_1 \preceq \ldots \preceq s'_r \in S',\ s'' \in S''\}$ is a basis of $\mathcal{V}$ and

$s'_1 \cdot ... \cdot s'_r s'' = \tilde{s}'_1 \cdot ... \cdot \tilde{s}'_{\tilde{r}} \tilde{s}'' \quad \Leftrightarrow \quad r = \tilde{r}, \quad s'_i = \tilde{s}'_i \ \forall i = 1, ..., r, \quad s'' = \tilde{s}''$;
ii) $(s')^{f'(s')} \in \mathcal{Z}(\mathcal{V}) \ \forall s' \in J'$ and $1 \in J'' \subseteq \mathcal{Z}(\mathcal{V})$;
iii) $1 \in T''$ and $\{s'_1 \cdot ... \cdot s'_r s'' | s'_1 \preceq ... \preceq s'_r \in T', \ s'' \in T''\}$ spans a subalgebra $\tilde{\mathcal{V}}$ of $\mathcal{V}$ (actually it is a basis of $\tilde{\mathcal{V}}$);
iv) if I is the ideal of $\mathcal{V}$ generated by $\{(s')^{f(s')} | s' \in J'\}$ then

$$\mathcal{Z}(\mathcal{V}/I) \cap \tilde{\mathcal{V}} \cong \oplus_{s'' \in J'' \cap T''} s''\mathbb{C}.$$

Then $\mathcal{Z}(\mathcal{V}) \cap \tilde{\mathcal{V}}$ is the subalgebra $\tilde{\mathcal{Z}}$ of $\tilde{\mathcal{V}}$ generated by

$$\{(s')^{f(s')}, s'' | s' \in J' \cap T', s'' \in J'' \cap T''\}.$$

Proof: Remark that obviously $\tilde{\mathcal{Z}} \subseteq \mathcal{Z}(\mathcal{V}) \cap \tilde{\mathcal{V}}$; moreover if $z, zx \in \mathcal{Z}(\mathcal{V})$ $(z \neq 0)$ then also $x \in \mathcal{Z}(\mathcal{V})$, because $\forall y \in \mathcal{V}$ $0 = [zx, y] = z[x, y]$, from which $[x, y] = 0$.
Let now $\prec'$ denote the linear ordering on the set of ordered monomials in the elements of S', which can be seen as $\oplus_{s' \in S'} \mathbb{N}$, defined by

$$(r_{s'})_{s' \in S'} \prec' (\tilde{r}_{s'})_{s' \in S'} \Leftrightarrow \exists s'_0 \in S' \text{ s. t. } r_{s'} = \tilde{r}_{s'} \ \forall s' \prec s'_0 \text{ and } r_{s'_0} < \tilde{r}_{s'_0}.$$

Notice that
1) if M and M' are ordered monomials in the elements of $\{(s')^{f(s')} | s' \in J'\}$ and of S' respectively then MM' is an ordered monomial in the elements of S';
2) if $M \prec' M'$ are ordered monomials in the elements of $\{(s')^{f(s')} | s' \in J'\}$ and $M'' = \prod (s')^{r_{s'}}$ is an ordered monomial in the elements of S' such that $r_{s'} < f(s') \ \forall s' \in J'$, then $\forall s'' \in S''$, $M' \nmid MM''s''$.
At this point let $z \in \mathcal{Z}(\mathcal{V}) \cap \tilde{\mathcal{V}} \setminus \tilde{\mathcal{Z}}$,

$$z = \sum_{M, M', s''} a_{M,M',s''} MM's''$$

where M is a monomial in the elements of $\{(s')^{f(s')} | s' \in J' \cap T'\}$,

$$M' = \prod (s')^{r_{s'}}$$

is an ordered monomial in the elements of T' such that $r_{s'} < f(s') \ \forall s' \in J'$, $s'' \in T''$ and $M's'' \notin J''$.
Let M_0 be $\prec'$-minimal in the set $\{M | a_{M,M',s''} \neq 0\}$ (which we suppose nonempty); then, thanks to iv),

$$x \doteq \sum_{M', s''} a_{M_0,M',s''} M's'' \in \tilde{\mathcal{V}}$$

is not central in $\mathcal{V}/I$, so that there exists $y\in\mathcal{V}$ such that $[x,y]\notin I$, that is there are an ordered monomial $\tilde{M}'=\prod(s')^{\tilde{r}_{s'}}$ such that $\tilde{r}_{s'}<f(s')\ \forall s'\in J'$ and an element $\tilde{s}''\in S''$ such that the coefficient c of $\tilde{M}'\tilde{s}''$ in $[x,y]$ is not zero.
But

$$0=[z,y]=\sum_{M,M',s''}a_{M,M',s''}M[M's'',y]=$$

$$=cM_0\tilde{M}'\tilde{s}''+M_0([x,y]-c\tilde{M}'\tilde{s}'')+\sum_{M\succ' M_0,M',s''}a_{M,M',s''}M[M's'',y],$$

which is impossible, thanks to 2).
Then $\mathcal{Z}(\mathcal{U})\subseteq\tilde{\mathcal{Z}}$. □

I shall now go deeper into the investigation of the relations between the multiplicity of ε in $\det H_\eta$ and the dimension of $\mathcal{Z}(\mathcal{U}_\varepsilon)\cap\mathcal{U}^+_{\varepsilon,\eta}$.

Lemma 2.
Let $\eta\in Q_+$. Suppose that $\{z_{\underline{\gamma}}\}_{\underline{\gamma}\in\mathrm{Par}(\eta)}\subset\tilde{\mathcal{U}}^-_{\mathbb{C}[q,q^{-1}],-\eta}$ is a linearly independent set which is mapped into a basis of $\mathcal{U}^-_{\varepsilon,-\eta}$ and suppose that $\forall\underline{\gamma}\in\mathrm{Par}(\eta)\ \exists m_{\underline{\gamma}}\in\mathbb{N}$ such that $(q-\varepsilon)^{m_{\underline{\gamma}}}|H_\eta(z_{\underline{\gamma}},x)\ \forall x\in\tilde{\mathcal{U}}^-_{\mathbb{C}[q,q^{-1}]}$; then $(q-\varepsilon)^{\sum_{\underline{\gamma}\in\mathrm{Par}(\eta)}m_{\underline{\gamma}}}|\det H_\eta$.

Proof: Let $\{x_{\underline{\gamma}}|\underline{\gamma}\in\mathrm{Par}(\eta)\}$ be a $\mathbb{C}[q,q^{-1}]$-basis of $\tilde{\mathcal{U}}^-_{\mathbb{C}[q,q^{-1}],-\eta}$ and A be the matrix (with coefficients in $\mathbb{C}[q,q^{-1}]$) sending $\underline{x}$ to $\underline{z}$; then the specialization A_ε of A is the matrix (with coefficients in $\mathbb{C}$) of change of basis from $\underline{x}$ to $\underline{z}$ in $\mathcal{U}^-_{\varepsilon,-\eta}$, hence $\det A_\varepsilon\neq 0$; but $\det A_\varepsilon=(\det A)_\varepsilon$, hence $(q-\varepsilon)\nmid\det A$.
In particular the multiplicity of ε in $\det H_\eta$ is the same as the multiplicity of ε in $\det H_\eta(\underline{z},\underline{x})$. But $H_\eta(\underline{z},\underline{x})$ is a $\mathrm{par}(\eta)\times\mathrm{par}(\eta)$ matrix whose $\underline{\gamma}^{\mathrm{th}}$ row is a multiple of $(q-\varepsilon)^{m_{\underline{\gamma}}}\ \forall\underline{\gamma}\in\mathrm{Par}(\eta)$, hence $(q-\varepsilon)^{\sum_{\underline{\gamma}\in\mathrm{Par}(\eta)}m_{\underline{\gamma}}}$ divides $\det H_\eta$.
□

Lemma 3.
Suppose given $z_i\in\tilde{\mathcal{U}}^+_{\mathbb{C}[q,q^{-1}],\gamma_i}$ $(i=1,\dots,m)$ with $\gamma_i>0$ and $w_i\in\tilde{\mathcal{U}}^+_{\mathbb{C}[q,q^{-1}],\tilde{\gamma}_i}$ $(i=0,\dots,m)$ with $\tilde{\gamma}_i\geq 0$, and suppose that $\forall i=1,\dots,m\ \forall x\in\tilde{\mathcal{U}}^-_{\mathbb{C}[q,q^{-1}]}$ there exist $y\in\tilde{\mathcal{U}}_{\mathbb{C}[q,q^{-1}]}$ such that

$$z_ix-(q-\varepsilon)y\in\bigoplus_{\alpha>0}\tilde{\mathcal{U}}_{\mathbb{C}[q,q^{-1}]}\tilde{\mathcal{U}}^+_{\mathbb{C}[q,q^{-1}],\alpha}$$

(note that this in particular happens when $z_i\in\mathcal{Z}(\mathcal{U}_\varepsilon)\cap\mathcal{U}^+_{\varepsilon,\gamma_i}$).
Then $\forall x\in\tilde{\mathcal{U}}^-_{\mathbb{C}[q,q^{-1}]}\ \exists y\in\tilde{\mathcal{U}}_{\mathbb{C}[q,q^{-1}]}$ such that

$$w_0z_1w_1\cdot\ldots\cdot z_mw_mx-(q-\varepsilon)^my\in\bigoplus_{\alpha>0}\tilde{\mathcal{U}}_{\mathbb{C}[q,q^{-1}]}\tilde{\mathcal{U}}^+_{\mathbb{C}[q,q^{-1}],\alpha}.$$

In particular if $\eta \doteq \sum_{i=1}^{m} \gamma_i + \sum_{i=0}^{m} \tilde{\gamma}_i$, then $\forall x \in \tilde{\mathcal{U}}^-_{\mathbb{C}[q,q^{-1}],-\eta}$

$$(q-\varepsilon)^m | \pi(w_0 z_1 w_1 \cdot \ldots \cdot z_m w_m x) = H_\eta(\Omega(w_0 z_1 \cdot \ldots \cdot z_m w_m), x)$$

where π is as defined in definition 1.B1.3.

Proof: Induction on m, the case $m = 0$ being obvious.
If $m > 0$, the inductive hypothesis implies that $\forall x \in \tilde{\mathcal{U}}^-_{\mathbb{C}[q,q^{-1}]}$

$$w_0 \cdot \ldots \cdot z_m w_m x = w_0 z_1 \big((q-\varepsilon)^{m-1} u + \tilde{u}\big)$$

with $u \in \mathcal{U}_q$, $\tilde{u} \in \bigoplus_{\alpha>0} \tilde{\mathcal{U}}_{\mathbb{C}[q,q^{-1}]} \tilde{\mathcal{U}}^+_{\mathbb{C}[q,q^{-1}],\alpha}$; of course it can be supposed that

$$u = \sum_{\lambda \in Q} u_\lambda K_\lambda \text{ with } u_\lambda \in \tilde{\mathcal{U}}^-_{\mathbb{C}[q,q^{-1}]} \quad \forall \lambda \in Q.$$

Hence I'm reduced to prove that $\forall \lambda \in Q$ $\forall x \in \tilde{\mathcal{U}}^-_{\mathbb{C}[q,q^{-1}]}$ $w_0 z_1 x K_\lambda = (q-\varepsilon)u + \tilde{u}$ with

$$u \in \tilde{\mathcal{U}}_{\mathbb{C}[q,q^{-1}]}, \quad \tilde{u} \in \bigoplus_{\alpha>0} \tilde{\mathcal{U}}_{\mathbb{C}[q,q^{-1}]} \tilde{\mathcal{U}}^+_{\mathbb{C}[q,q^{-1}],\alpha}:$$

indeed there exists $y \in \tilde{\mathcal{U}}_{\mathbb{C}[q,q^{-1}]}$ such that

$$w_0 z_1 x K_\lambda - (q-\varepsilon) w_0 y K_\lambda \in \bigoplus_{\alpha>0} \tilde{\mathcal{U}}_{\mathbb{C}[q,q^{-1}]} \tilde{\mathcal{U}}^+_{\mathbb{C}[q,q^{-1}],\alpha} K_\lambda$$

which is equal to

$$\bigoplus_{\alpha>0} \tilde{\mathcal{U}}_{\mathbb{C}[q,q^{-1}]} \tilde{\mathcal{U}}^+_{\mathbb{C}[q,q^{-1}],\alpha}.$$

□

Corollary 4.
Suppose given $\{x_\alpha | \alpha \in \tilde{R}_+\}$ with $x_\alpha \in \tilde{\mathcal{U}}^+_{\mathbb{C}[q,q^{-1}],\alpha}$ such that
1) $\{x(\underline{\gamma}) | \underline{\gamma} \in \cup_{\eta \in Q_+} \mathrm{Par}(\eta)\}$ is a basis of $\mathcal{U}^+_\varepsilon$;
2) $\exists f : J \to \mathbb{Z}_+$ $(J \subseteq \tilde{R}_+)$ such that $\forall \alpha \in J$ $\forall y \in \tilde{\mathcal{U}}^-_{\mathbb{C}[q,q^{-1}]}$ there exists $\tilde{y} \in \tilde{\mathcal{U}}_{\mathbb{C}[q,q^{-1}]}$ such that

$$x_\alpha^{f(\alpha)} y - (q-\varepsilon)\tilde{y} \in \bigoplus_{\alpha>0} \tilde{\mathcal{U}}_{\mathbb{C}[q,q^{-1}]} \tilde{\mathcal{U}}^+_{\mathbb{C}[q,q^{-1}],\alpha}$$

(this in particular happens if $x_\alpha^{f(\alpha)} \in \mathcal{Z}(\mathcal{U}_\varepsilon)$).
Then $\forall \eta \in Q_+$ the multiplicity of ε in $\det H_\eta$ is at least $\sum_{\substack{\alpha \in J \\ m>0}} \mathrm{par}\big(\eta - m f(\alpha) p(\alpha)\big)$.

Proof: Thanks to lemma 3.2.3, $\forall \underline{\gamma} \in \mathrm{Par}(\eta)$ $\forall y \in \tilde{\mathcal{U}}^-_{\mathbb{C}[q,q^{-1}],-\eta}$ we have that

$$(q-\varepsilon)^{\sum_{\alpha\in J}[\frac{\mu^\alpha_{\underline{\gamma}}}{f(\alpha)}]} \Big| \pi(x(\underline{\gamma})y),$$

where $\mu^\alpha_{\underline{\gamma}}$ denotes the multiplicity of α in $\underline{\gamma}$ (see definition 2.7.3).

Then lemma 3.2.2 implies that $(q-\varepsilon)^{\sum_{\underline{\gamma}\in\mathrm{Par}(\eta)}\sum_{\alpha\in J}[\frac{\mu^\alpha_{\underline{\gamma}}}{f(\alpha)}]} \Big| \det H_\eta$; but

$$\sum_{\underline{\gamma}\in\mathrm{Par}(\eta)}\sum_{\alpha\in J}\left[\frac{\mu^\alpha_{\underline{\gamma}}}{f(\alpha)}\right] =$$

$$= \sum_{\alpha\in J}\sum_{m>0} m\#\{\underline{\gamma}\in\mathrm{Par}(\eta) | mf(\alpha) \le \mu^\alpha_{\underline{\gamma}} < (m+1)f(\alpha)\} =$$

$$= \sum_{\alpha\in J}\sum_{m>0} m\big(\mathrm{par}(\eta - mf(\alpha)p(\alpha)) - \mathrm{par}(\eta-(m+1)f(\alpha)p(\alpha))\big) =$$

$$= \sum_{\substack{\alpha\in J\\ m>0}} \mathrm{par}\big(\eta - mf(\alpha,p(\alpha)\big),$$

which gives the thesis. □

Definition 5.
Let H_ε denote the contravariant form induced on the specialization $\mathcal{U}^-_\varepsilon$ at ε of $\mathcal{U}^-_q$ by H, that is given $x, y \in \tilde{\mathcal{U}}^-_{\mathbb{C}[q,q^{-1}]}$, if $\bar{x}, \bar{y}$ denote respectively the images of x and y in $\mathcal{U}^-_\varepsilon$, then $H_\varepsilon(\bar{x},\bar{y})$ is by definition the image in $\mathcal{U}^-_\varepsilon$ of $H(x,y)$.
Moreover $H_{\varepsilon,\eta}$ denotes the restriction of H_ε to $\mathcal{U}^-_{\varepsilon,-\eta}$.

Remark 6.
H_ε and $H_{\varepsilon,\eta}$ are well defined and $H_{\varepsilon,\eta}$ is the specialization at ε of H_η. □

Definition 7.
In the conditions of corollary 3.2.4, let $I^+_{\mathcal{Z}} \subseteq \mathcal{U}_\varepsilon$ be the $\mathbb{C}$-linear span of the set $\{x(\underline{\gamma}) | \exists \alpha \in J \text{ s.t. } \mu^\alpha_{\underline{\gamma}} \ge f(\alpha)\}$ and let $I^-_{\mathcal{Z}} \doteqdot \Omega(I^+_{\mathcal{Z}})$.

Remark 8.
Notice that $\mathcal{U}^+_\varepsilon / I^+_{\mathcal{Z}}$ is a Q-graded $\mathbb{C}$-vector space with basis

$$\{x(\underline{\gamma}) | \mu^\alpha_{\underline{\gamma}} < f(\alpha) \forall \alpha \in J\}$$

and that $I^-_{\mathcal{Z}} \subseteq \ker H_\varepsilon$, so that H_ε induces on $\mathcal{U}^-_\varepsilon / I^-_{\mathcal{Z}}$ a Hermitian form (again denoted by H_ε).

Proof: Everything follows from lemma 3.2.3. □

Remark 9.
In the conditions of corollary 3.2.4, suppose that $x_\alpha^{f(\alpha)} \in \mathcal{Z}(\mathcal{U}_\varepsilon)$ $\forall \alpha \in J$; then
1) $I_{\mathcal{Z}}^-$ and $I_{\mathcal{Z}}^+$ are ideals of $\mathcal{U}_\varepsilon^-$ and $\mathcal{U}_\varepsilon^+$ respectively;
2) the ideal $I_{\mathcal{Z}}$ of $\mathcal{U}_\varepsilon$ generated by $I_{\mathcal{Z}}^-$ and $I_{\mathcal{Z}}^+$ is $I_{\mathcal{Z}} = I_{\mathcal{Z}}^- \mathcal{U}_\varepsilon^{\geq 0} + \mathcal{U}_\varepsilon^{\leq 0} I_{\mathcal{Z}}^+$;
3) $\mathcal{U}_\varepsilon / I_{\mathcal{Z}} \cong \mathcal{U}_\varepsilon^- / I_{\mathcal{Z}}^- \otimes_{\mathbb{C}} \mathcal{U}_\varepsilon^0 \otimes_{\mathbb{C}} \mathcal{U}_\varepsilon^+ / I_{\mathcal{Z}}^+$. This means that one can define $(\mathcal{U}_\varepsilon / I_{\mathcal{Z}})^-$, $(\mathcal{U}_\varepsilon / I_{\mathcal{Z}})^0$, $(\mathcal{U}_\varepsilon / I_{\mathcal{Z}})^+$, and $(\mathcal{U}_\varepsilon / I_{\mathcal{Z}})^- \cong \mathcal{U}_\varepsilon^- / I_{\mathcal{Z}}^-$, $(\mathcal{U}_\varepsilon / I_{\mathcal{Z}})^0 \cong \mathcal{U}_\varepsilon^0$;
4) π is well defined on $\mathcal{U}_\varepsilon / I_{\mathcal{Z}}$ and $H_\varepsilon(x,y) = \pi(\Omega(x)y)$ $\forall x, y \in \mathcal{U}_\varepsilon^- / I_{\mathcal{Z}}^-$; in particular H_ε is a contravariant form.

Proof: The claim is obvious. □

Lemma 10.
In the conditions of corollary 3.2.4, if $\forall \eta$ the multiplicity of ε in $\det H_\eta$ is exactly $\sum_{\substack{\alpha \in J \\ m>0}} \mathrm{par}\big(\eta - mf(\alpha)p(\alpha)\big)$, then $I_{\mathcal{Z}}^- = \ker H_\varepsilon$, so that $I_{\mathcal{Z}}^-$ is an ideal of $\mathcal{U}_\varepsilon^-$ and $\mathcal{U}_\varepsilon^- / I_{\mathcal{Z}}^-$ is an algebra. Morover H_ε is non degenerate on $\mathcal{U}_\varepsilon^- / I_{\mathcal{Z}}^-$, that is $\forall x \neq 0$ in $\mathcal{U}_\varepsilon^- / I_{\mathcal{Z}}^-$ there exists $y \in \mathcal{U}_\varepsilon^- / I_{\mathcal{Z}}^-$ such that $H_\varepsilon(x,y) \neq 0$.

Proof: I prove that $I_{\mathcal{Z}}^- = \ker H_\varepsilon$, from which everything follows immediately (because the kernel of a contrava⁻iant form is an ideal).
Remark that $\ker H_\varepsilon$ is a graded ideal of $\mathcal{U}_\varepsilon^-$ and that $\ker H_\varepsilon \cap \mathcal{U}_{\varepsilon,0}^- = \{0\}$, and suppose that the image in $\mathcal{U}_\varepsilon^-$ of $x = \sum_{\underline{\gamma} \in \mathrm{Par}(\eta)} a_{\underline{\gamma}} \Omega(x(\underline{\gamma})) \in \tilde{\mathcal{U}}_{\mathbb{C}[q,q^{-1}],\eta}^-$ $(\eta > 0)$ is in $\ker H_\varepsilon \setminus I_{\mathcal{Z}}^-$; then there exists $\underline{\gamma} \in \mathrm{Par}(\eta)$ such that $a_{\underline{\gamma}} \neq 0$ in $\mathcal{U}_\varepsilon^-$ and, since $x \notin I_{\mathcal{Z}}^-$, I can suppose that $\mu_{\underline{\gamma}}^\alpha < f(\alpha)$ $\forall \alpha \in J$.
This implies that $\{x(\underline{\tilde{\gamma}}) | \underline{\tilde{\gamma}} \neq \underline{\gamma}\} \cup \{\Omega(x)\}$ is a basis of $\mathcal{U}_\varepsilon^+$ and, using lemma 3.2.2 and lemma 3.2.3, one has that

$$(q-\varepsilon)^{1+\sum_{\underline{\tilde{\gamma}} \in \mathrm{Par}(\eta) \setminus \{\underline{\gamma}\}} \sum_{\alpha \in J} [\frac{\mu_{\underline{\tilde{\gamma}}}^\alpha}{\alpha}]} \,|\, \det H_\eta.$$

Since $[\frac{\mu_{\underline{\gamma}}^\alpha}{\alpha}] = 0$ $\forall \alpha \in J$, this says that the multiplicity of ε in $\det H_\eta$ is at least $1 + \sum_{\substack{\alpha \in J \\ m>0}} \mathrm{par}(\eta - mf(\alpha)p(\alpha))$, contradicting the hypothesis. Then $\ker H_\varepsilon \subseteq I_{\mathcal{Z}}^-$ and, thanks to remark 3.2.8, $I_{\mathcal{Z}}^- = \ker H_\varepsilon$. □

Lemma 11.
In the hypotheses of remark 3.2.9 and of lemma 3.2.10

$$\mathcal{Z}(\mathcal{U}_\varepsilon / I_{\mathcal{Z}}) \cap \mathcal{U}_\varepsilon^- / I_{\mathcal{Z}}^- = \mathbb{C} = \mathcal{Z}(\mathcal{U}_\varepsilon / I_{\mathcal{Z}}) \cap \mathcal{U}_\varepsilon^+ / I_{\mathcal{Z}}^+.$$

Proof: Indeed let $x \in \mathcal{U}_{\varepsilon,-\alpha}^-$ with $\alpha > 0$ be such that $x + I_{\mathcal{Z}} \in \mathcal{Z}(\mathcal{U}_\varepsilon / I_{\mathcal{Z}})$ and let $y \in \mathcal{U}_\varepsilon^-$; then $H_\varepsilon(x,y) = \pi(\Omega(x)y) = \pi(y\Omega(x)) = 0$. Thus

$$\mathcal{Z}(\mathcal{U}_\varepsilon / I_{\mathcal{Z}}) \cap \mathcal{U}_\varepsilon^- / I_{\mathcal{Z}}^- \subseteq \ker H_\varepsilon \oplus \mathbb{C} = \mathbb{C} \subseteq \mathcal{Z}(\mathcal{U}_\varepsilon / I_{\mathcal{Z}}) \cap \mathcal{U}_\varepsilon^- / I_{\mathcal{Z}}^-.$$

On the other hand Ω is an antiautomorphism of $\mathcal{U}_\varepsilon / I_{\mathcal{Z}}$ mapping $\mathcal{U}_\varepsilon^+ / I_{\mathcal{Z}}^+$ isomorphically onto $\mathcal{U}_\varepsilon^- / I_{\mathcal{Z}}^-$, and the claim follows. □

Proposition 12.
In the hypotheses of remark 3.2.9 and of lemma 3.2.10, $\mathcal{Z}(\mathcal{U}_\varepsilon)\cap\mathcal{U}_\varepsilon^+$ is the $\mathbb{C}$-linear span of

$$\{x(\underline{\gamma})|\forall\alpha\in J\ f(\alpha)|\mu_{\underline{\gamma}}^\alpha \text{ and } \forall\alpha\notin J\ \mu_{\underline{\gamma}}^\alpha=0\}.$$

This means that $\mathcal{Z}(\mathcal{U}_\varepsilon)\cap\mathcal{U}_\varepsilon^+$ is the algebra of polynomials in $\{x_\alpha^{f(\alpha)}|\alpha\in J\}$.
Proof: It follows applying lemma 3.2.1, setting:

$$\mathcal{V}\doteqdot\mathcal{U}_\varepsilon,\quad \tilde{\mathcal{V}}\doteqdot\mathcal{U}_\varepsilon^+,\quad I\doteqdot I_{\mathcal{Z}},$$

$$T'\doteqdot\{x_\alpha|\alpha\in\tilde{R}_+\},\quad S'\doteqdot T'\cup\Omega(T'),\quad J'\doteqdot\{x_\alpha,\Omega(x_\alpha)|\alpha\in J\},$$
$$J''=T''\doteqdot\{1\},\quad S''\doteqdot\{K_\lambda|\lambda\in Q\},$$
$$f'(x_\alpha)\doteqdot f(\alpha),\quad f'(\Omega(x_\alpha))\doteqdot f(\alpha)\quad\forall\alpha\in J,$$

and using lemma 3.2.11. □

§3. Some central vectors: real case.

The next step is to exhibit explicitely some central elements in $\mathcal{U}_\varepsilon$ where ε is a primitive l^{th} root of 1.

Remark 1.
K_δ is central in $\mathcal{U}_q$, and henceforth $K_\delta\in\mathcal{Z}(\mathcal{U}_\varepsilon)$.

Proposition 2.
$\forall i\in I\ E_i^{l_i}\in\mathcal{Z}(\mathcal{U}_\varepsilon)$.
Proof: The following proof is due to Kac:

$$E_i^{l_i}K_\lambda=q^{-l_i(\alpha_i|\lambda)}K_\lambda E_i^{l_i}=K_\lambda E_i^{l_i}\ \text{ in } \mathcal{U}_\varepsilon\ \ \forall\lambda\in Q.$$

$$E_i^{l_i}F_j=F_jE_i^{l_i}\quad\forall j\neq i\ (j\in I);$$

$$E_i^{l_i}F_i=F_iE_i^{l_i}+[l_i]_{q_i}\frac{q_i^{1-l_i}K_i-q_i^{l_i-1}K_i^{-1}}{q_i-q_i^{-1}}E_i^{l_i-1}=$$
$$=F_iE_i^{l_i}\ \text{ in } \mathcal{U}_\varepsilon.$$

Of course $E_i^{l_i}$ commutes with E_i.
Let $j\in I$ be different from i. Recall (see proposition 1.B3.8) that

$$\text{ad}E_i^m(E_j)=0\ \text{ when }\ m\geq 1-a_{i,j}.$$

Remark that the hypotheses on l imply that $l_i=l\geq 1-a_{i,j}$ whenever $\mathfrak{g}\neq G_2$, or $3\nmid l$, or $d_i\neq 3$, while if $\mathfrak{g}=G_2$ and $d_i=3|l$ we have that $a_{i,j}=1$

and $l_i = \frac{l}{3} > 1 = -a_{i,j}$ because $l > 3$, so that again $l_i \geq 1 - a_{i,j}$. Then $\mathrm{ad}E_i^{l_i}(E_j) = 0$.
I want now to prove that $\mathrm{ad}E_i^{l_i}(E_j) = [E_i^{l_i}, E_j]$: indeed (see lemma 1.B3.5)

$$\Delta(E_i^{l_i}) = \sum_{r=0}^{l_i} q^{-r(l_i-r)} \begin{bmatrix} l_i \\ r \end{bmatrix}_{q_i} K_i^{l_i-r} E_i^r \otimes E_i^{l_i-r};$$

But $\begin{bmatrix} l_i \\ r \end{bmatrix}_{q_i} = 0$ in $\mathcal{U}_\varepsilon$ whenever $0 < r < l_i$, while $\begin{bmatrix} l_i \\ r \end{bmatrix}_{q_i} = 1$ if $r = 0, l_i$, then $\Delta(E_i^{l_i}) = E_i^{l_i} \otimes 1 + K_i^{l_i} \otimes E_i^{l_i}$.
Since $K_i^{l_i}$ is central in $\mathcal{U}_\varepsilon$ and

$$K_i^{l_i} S(E_i^{l_i}) = K_i^{l_i}(-K_i^{-1}E_i)^l = -q_i^{l_i(l_i-1)} E_i^{l_i} = -E_i^{l_i} \text{ in } \mathcal{U}_\varepsilon,$$

we have that

$$\mathrm{ad}E_i^{l_i}(E_j) = E_i^{l_i} E_j - E_j E_i^{l_i} = [E_i^{l_i}, E_j],$$

from which the claim follows. □

Corollary 3.
$\forall \alpha \in R_+^{\mathrm{re}}\ E_\alpha^{l_\alpha} \in \mathcal{Z}(\mathcal{U}_\varepsilon)$.
Proof: The thesis immediately follows from the fact $\forall i \in I\ T_i$ is an automorphism of $\mathcal{U}_\varepsilon$ and from the definition of E_α and d_α, thanks to proposition 3.3.2. □

In this way I have found a central vector of $\mathcal{U}_\varepsilon$ for every positive real root α.

§4. Some central vectors: imaginary case.

In order to exhibit central vectors for the imaginary roots, I need the following lemmas:

Lemma 1.
Let $x \in \tilde{\mathcal{U}}_{\mathbb{C}[q,q^{-1}]}$ be such that $\forall i \in I_0$
1) $T_{\omega_i^-}(x) = x$
2) $[x, E_i] = [x, F_i] = [x, K_i] = 0$ in $\mathcal{U}_\varepsilon$;
then $x \in \mathcal{Z}(\mathcal{U}_\varepsilon)$.
Proof: Notice first that the hypotheses imply that

$$[x, K_0] = [x, K_\delta K_\theta^{-1}] = [x, K_\delta]K_\theta^{-1} + K_\delta[x, K_\theta^{-1}] = 0$$

since K_θ is a monomial in $\{K_i | i \in I_0\}$ and K_δ is central.
Moreover, if $x \in \tilde{\mathcal{U}}_{\mathbb{C}[q,q^{-1}]}$ satisfies the hypotheses, also $\Omega(x)$ does, and

$$[\Omega(x), E_0] = 0 \text{ in } \mathcal{U}_\varepsilon \Rightarrow [x, F_0] = 0 \text{ in } \mathcal{U}_\varepsilon.$$

Hence it is enough to prove that if x satisfies the conditions of the lemma then $[x, E_0] = 0$ in $\mathcal{U}_\varepsilon$.
I distinguish two cases:
a) $\exists i \in I_0$ such that $\langle \check{\omega_i}, \theta \rangle = 1$, that is $\theta - 2\alpha_i \notin Q_+$ (this happens when $\mathfrak{g} \neq E_8, F_4, G_2$).
Notice that

$$(\check{\omega_i})^{-1} s_0(\alpha_0) = -(\check{\omega_i})^{-1}(\alpha_0) = (\check{\omega_i})^{-1}(\theta - \delta) = \theta + \langle \check{\omega_i}, \theta \rangle \delta - \delta = \theta;$$

moreover $l(s_0\check{\omega_i}) < l(\check{\omega_i})$, hence, thanks to remark 1.B2.8, we have that

$$T^{-1}_{s_0\check{\omega_i}}(F_0) \in \mathcal{U}^-_{q,-(\check{\omega_i})^{-1}s_0(\alpha_0)} = \mathcal{U}^-_{q,-\theta}$$

which is contained in the subalgebra of $\tilde{\mathcal{U}}_{\mathbb{C}[q,q^{-1}]}$ generated by $\{F_j | j \in I_0\}$.
This implies that $[x, T^{-1}_{s_0\check{\omega_i}}(F_0)] = 0$, hence

$$[x, T_0(F_0)] = T_{\check{\omega_i}}[x, T^{-1}_{\check{\omega_i}} T_0(F_0)] = T_{\check{\omega_i}}[x, T^{-1}_{s_0\check{\omega_i}}(F_0)] = 0.$$

But $T_0(F_0) = -K_0^{-1}E_0$, so that

$$[x, E_0] = -[x, K_0 T_0(F_0)] = -K_0[x, T_0(F_0)] = 0.$$

b) $\mathfrak{g} = E_8, F_4, G_2$; in this case $\not\exists i \in I_0$ such that $\langle \check{\omega_i}, \theta \rangle = 1$, but $\langle \check{\omega_n}, \theta \rangle = 2$, $\langle \check{\omega}_{n-1}, \theta \rangle = 3$, $a_{n,n-1} = -1$, $a_{n,0} = a_{0,n} = -1$ and $a_{n,r} = 0$ if $1 \leq r < n-1$. Set $\tilde{\theta} \doteq s_n(\theta) = \theta - \alpha_n = \alpha_n + 3\alpha_{n-1} + \sum_{1 \leq r < n-1} n_r \alpha_r$, so that $\langle \check{\omega_n}, \tilde{\theta} \rangle = 1$ and $\check{\omega_n}(\tilde{\theta}) = \tilde{\theta} - \delta = \theta - \alpha_n - \delta = -\alpha_n - \alpha_0$, that is $\tilde{\theta} = (\check{\omega_n})^{-1}(-\alpha_0 - \alpha_n) =$ $= (\check{\omega_n})^{-1} s_0(-\alpha_n)$; it follows that $l((\check{\omega_n})^{-1} s_0 s_n) < l((\check{\omega_n})^{-1} s_0) < l((\check{\omega_n})^{-1})$; in particular $T^{-1}_{s_n s_0 \check{\omega_n}}(F_n) \in \mathcal{U}^-_{q,-(\check{\omega_n})^{-1}s_0 s_n(\alpha_n)} = \mathcal{U}^-_{q,-\tilde{\theta}}$ which is contained in the subalgebra of $\tilde{\mathcal{U}}_{\mathbb{C}[q,q^{-1}]}$ generated by $\{F_j | j \in I_0\}$.
Again this implies $[x, T^{-1}_{s_n s_0 \check{\omega_n}}(F_n)] = 0$ so that

$$[x, T_0 T_n(F_n)] = T_{\check{\omega_n}}[x, T^{-1}_{s_n s_0 \check{\omega_n}}(F_n)] = 0.$$

But $T_0 T_n(F_n) = -T_0(K_n^{-1}E_n) = -K_{-(\alpha_0+\alpha_n)} T_0(E_n)$ and

$$[x, T_0(E_n)] = -K_{\alpha_0+\alpha_n}[x, T_0 T_n(F_n)] = 0.$$

In particular

$$0 = [x, [T_0(E_n), F_n]] = -[x, K_n E_0] = -K_n[x, E_0],$$

from which $[x, E_0] = 0$. □

Lemma 2.
Fix $m > 0$ and let x be a linear combination (with coefficients in $\mathbb{C}[q,q^{-1}]$) of the vectors $E_{(m\delta,i)}$ for $i \in I_0$, that is $x = \sum_{i\in I_0} b_i E_{(m\delta,i)}$. Then x is central in $\mathcal{U}_\varepsilon$ if and only if $\forall j \in I_0$ $(q-\varepsilon)$ divides the element $\bar{b}_j$ where

$$\bar{b}_j \doteq \sum_{i\in I_0} o(i)^m [ma_{i,j}]_{q_i} b_i = [m]_q \sum_{i\in I_0} o(i)^m \frac{[(\alpha_i|\alpha_j)]_{q^m}}{[d_i]_q} b_i.$$

Proof: Since x is T_{ω_i}-stable and $[x, K_i] = 0$ $\forall i \in I_0$, lemma 3.4.1 implies that it is enough to prove that $\forall j \in I$ $(q-\varepsilon)|\bar{b}_j$ if and only if $[x, E_j] = [x, F_j] = 0$ in $\mathcal{U}_\varepsilon$, or, which is equivalent, $[x, E_j] = [x, -K_j^{-1}F_j] = 0$ in $\mathcal{U}_\varepsilon$.
Now, using proposition 2.4.15, we see that

$$[x, E_j] = \sum_{i\in I_0} b_i [E_{(m\delta,i)}, E_j] =$$

$$= \frac{o(j)^m}{m} \sum_{i\in I_0} o(i)^m [ma_{i,j}]_{q_i} b_i T_{\omega_j}^{-m}(E_j) = \frac{o(j)^m \bar{b}_j}{m} T_{\omega_j}^{-m}(E_j),$$

and analogously

$$[x, -K_j^{-1}F_j] = \sum_{i\in I_0} b_i [E_{(m\delta,i)}, -K_j^{-1}F_j] =$$

$$= -\frac{o(j)^m}{m} \sum_{i\in I_0} o(i)^m [ma_{i,j}]_{q_i} b_i T_{\omega_j}^{m}(-K_j^{-1}F_j) = -\frac{o(j)^m \bar{b}_j}{m} T_{\omega_j}^{m}(-K_j^{-1}F_j).$$

The claim then immediately follows. □

Remark 3.
In the hypotheses of lemma 3.4.2, $(q-\varepsilon)|\bar{b}_j$ if and only if $[x, F_{(m\delta,j)}] = 0$ in $\mathcal{U}_\varepsilon$ (or, which is equivalent, $[E_{(m\delta,j)}, \Omega(x)] = 0$ in $\mathcal{U}_\varepsilon$).
Proof: It follows from the fact that, thanks to proposition 2.4.17,

$$[x, F_{(m\delta,j)}] = \frac{o(j)^m}{m} \sum_{i\in I_0} o(i)^m [ma_{i,j}]_{q_i} b_i \frac{K_{r\delta} - K_{-r\delta}}{q_j - q_j^{-1}} =$$

$$= \frac{o(j)^m}{m} \bar{b}_j \frac{K_{r\delta} - K_{-r\delta}}{q_j - q_j^{-1}}.$$

□

Proposition 4.
$\forall m > 0$ $\forall i \in I_0$ $E_{(l_i m\delta,i)} \in \mathcal{Z}(\mathcal{U}_\varepsilon)$.

Proof: Thanks to lemma 3.4.2

$$E_{(l_i m\delta,i)} \in \mathcal{Z}(\mathcal{U}_\varepsilon) \text{ if and only if } (q-\varepsilon)|[ml_i a_{i,j}]_{q_i} \quad \forall j \in I_0;$$

but $l|2ml_i d_i a_{i,j}$ $\forall m > 0$, so the claim follows. □

Definition 5.
Suppose $\mathfrak{g} = A_n$; then if $\frac{l}{\text{g.c.d.}(l, n+1)}|m$ but $l \nmid m$ I define

$$E_{m\delta} \doteqdot \Omega(\bar{F}_{(m\delta,1)}),$$

(see proposition 2.5.6).

Proposition 6.
If $\mathfrak{g} = A_n$ and $\frac{l}{\text{g.c.d.}(l, n+1)}|m$ but $l \nmid m$, $E_{m\delta} \in \mathcal{Z}(\mathcal{U}_\varepsilon) \setminus \{0\}$.

Proof: First notice that $E_{m\delta} \neq 0$ because $A'^{(m)}_{1,n} = 1 \neq 0$.
If $j > 1$ the definition of $\bar{F}_{(m\delta,1)}$ implies that

$$[E_{(m\delta,j)}, \Omega(E_{m\delta})] = [E_{(m\delta,j)}, \bar{F}_{(m\delta,1)}] = 0.$$

On the other hand in the notations of lemma 3.4.2

$$\bar{b}_1 = [m]_q \sum_{i \in I_0} \frac{o(i)^m [(\alpha_1|\alpha_i)]_{q^m}}{[d_i]_q} o(i)^m [d_i]_q A'^{(m)}_{1,i} = [m]_q \bar{c}^{(m)}_1 = [m]_q [n+1]_{q^m}$$

which is zero in $\mathcal{U}_\varepsilon$ because $l|m(n+1)$ and $l \nmid m$.
This is the claim, thanks lemma 3.4.2 and remark 3.4.3. □

Definition 7.
Suppose $\mathfrak{g} = E_6$ and $3|l$; then if $\frac{l}{3}|m$ but $l \nmid m$ I define $E_{m\delta} \doteqdot \Omega(\bar{F}_{(m\delta,2)})$.

Proposition 8.
If $\mathfrak{g} = E_6$, $3|l$ and $\frac{l}{3}|m$ but $l \nmid m$, then $E_{m\delta} \in \mathcal{Z}(\mathcal{U}_\varepsilon) \setminus \{0\}$.

Proof: Since $A'^{(m)}_{2,6} = 1 \neq 0$, $E_{m\delta} \neq 0$.
If $j > 2$ the definition of $\bar{F}_{(m\delta,2)}$ implies that

$$[E_{(m\delta,j)}, \Omega(E_{m\delta})] = [E_{(m\delta,j)}, \bar{F}_{(m\delta,2)}] = 0.$$

On the other hand in the notations of lemma 3.4.2

$$\bar{b}_2 = [m]_q \sum_{i \geq 2} \frac{o(i)^m [(\alpha_2|\alpha_i)]_{q^m}}{[d_i]_q} o(i)^m [d_i]_q A'^{(m)}_{2,i} =$$

$$= [m]_q \bar{c}^{(m)}_2 = [m]_q A'^{(m)}_{1,1} = [m]_q [6]_{q^m}$$

which is zero in $\mathcal{U}_\varepsilon$ because $l|6m$.
Finally $\bar{b}_1 = [m]_q \sum_{i\geq 2}[(\alpha_i|\alpha_1)]_{q^m} A_{2,i}'^{(m)} = -[m]_q A_{2,4}'^{(m)} = -[m]_q[3]_{q^m} = 0$ in $\mathcal{U}_\varepsilon$ because $l|3m$.
The claim is thus proved. □

Remark 9.
The conditions given in definitions 3.4.5 and 3.4.7 in order to define $E_{m\delta}$ are equivalent to the condition $\bar{c}_1^{(m)} = 0$ in $\mathcal{U}_\varepsilon$ (see definition 2.6.6). □

§5. The center of $\mathcal{U}_\varepsilon$.

In this section I use the results obtained till now to describe the center of $\mathcal{U}_\varepsilon$. I start investigating the positive part of this center.

Theorem 1.
$\mathcal{Z}(\mathcal{U}_\varepsilon) \cap \mathcal{U}_\varepsilon^+$ is the algebra of polynomials

$$\mathbb{C}[E_\alpha^{l_\alpha}, E_{(l_i m\delta, i)}, E_{r\delta}]$$

where $\alpha \in R_+^{\text{re}}$, $m, r > 0$, $i \in I_0$ and $\bar{c}_1^{(r)} = 0$ in $\mathcal{U}_\varepsilon$.
In particular, if $\bar{c}_1^{(m)} \neq 0$ in $\mathcal{U}_\varepsilon$ $\forall m > 0$, then

$$\mathcal{Z}(\mathcal{U}_\varepsilon) \cap \mathcal{U}_\varepsilon^+ = \mathbb{C}[E_\alpha^{l_\alpha}, E_{(l_i m\delta, i)} | \alpha \in R_+^{\text{re}}, m > 0, i \in I_0];$$

if $\mathfrak{g} = A_n$ and g.c.d.$(l, n+1) \neq 1$,

$$\mathcal{Z}(\mathcal{U}_\varepsilon) \cap \mathcal{U}_\varepsilon^+ =$$

$$= \mathbb{C}[E_\alpha^l, E_{(lm\delta, i)}, E_{\frac{l}{\text{g.c.d.}(l,n+1)} r\delta} | \alpha \in R_+^{\text{re}}, m, r > 0, i \in I_0, (l, n+1) \nmid r];$$

and finally if $\mathfrak{g} = E_6$, $3|l$,

$$\mathcal{Z}(\mathcal{U}_\varepsilon) \cap \mathcal{U}_\varepsilon^+ = \mathbb{C}[E_\alpha^l, E_{(lm\delta, i)}, E_{\frac{l}{3} r\delta} | \alpha \in R_+^{\text{re}}, m, r > 0, i \in I_0, 3 \nmid r].$$

Proof: Set

$$J_0 \doteqdot \{(m\delta, n) \in \tilde{R}_+ | \bar{c}_1^{(m)} = 0 \text{ in } \mathcal{U}_\varepsilon\}$$

$$\forall \alpha \in \tilde{R}_+ \quad x_\alpha \doteqdot \begin{cases} E_{p(\alpha)} & \text{if } \alpha \in J_0 \\ E_\alpha & \text{otherwise;} \end{cases}$$

$$\tilde{R}_{+,l} \doteqdot R_+^{\text{re}} \cup \{(l_i m\delta, i) | i \in I_0, m > 0\};$$

$$J \doteqdot \tilde{R}_{+,l} \cup J_0, \quad \text{and}$$

$$\forall \alpha \in J \quad f(\alpha) = \begin{cases} l_\alpha & \text{if } \alpha \in R_+^{\text{re}} \\ 1 & \text{otherwise.} \end{cases}$$

The claim is that $\mathcal{Z}(\mathcal{U}_\varepsilon) \cap \mathcal{U}_\varepsilon^+ = \mathbb{C}[x_\alpha^{f(\alpha)} | \alpha \in J]$.
Of course $x_\alpha^{f(\alpha)} \in \mathcal{Z}(\mathcal{U}_\varepsilon) \cap \mathcal{U}_\varepsilon^+ \ \forall \alpha \in J$.
On the other hand notice that $\forall m > 0$ $\{E_{(m\delta,i)} | i \in I_0\}$ and $\{x_{(m\delta,i)} | i \in I_0\}$ span the same $\mathbb{C}$-vector space: indeed

$$x_{(m\delta,i)} = \begin{cases} E_{(m\delta,i)} & \text{if } \bar{c}_1^{(m)} \neq 0 \text{ in } \mathcal{U}_\varepsilon \text{ or } i \neq n \\ E_{m\delta} = E_{(m\delta,n)} + \sum_{j \neq n} b_j E_{(m\delta,j)} & \text{if } \bar{c}_1^{(m)} = 0 \text{ in } \mathcal{U}_\varepsilon \text{ and } i = n; \end{cases}$$

this implies that the conditions of corollary 3.2.4 hold, so that the multiplicity of ε in $\det H_\eta$ is at least

$$\sum_{\substack{\alpha \in J \\ m > 0}} \mathrm{par}(\eta - mf(\alpha)p(\alpha)) = \sum_{\substack{\alpha \in \tilde{R}_+ \\ m > 0}} \mathrm{par}(\eta - ml_\alpha p(\alpha)) + \sum_{\substack{(m\delta,n) \in J_0 \\ r > 0}} \mathrm{par}(\eta - mr\delta);$$

but theorem 3.1.9 implies that this is exactly the multiplicity of ε in $\det H_\eta$, hence proposition 3.2.12 applies, and the claim follows. □

Lemma 2.
$\mathcal{U}_{\varepsilon,\mathrm{im}}^+ \cap \mathcal{Z}(\mathcal{U}_\varepsilon^+) \subseteq \mathcal{Z}(\mathcal{U}_\varepsilon)$.

Proof: First notice that $\forall i \in I_0$ T_{ω_i} acts as the identity on $\mathcal{U}_{\varepsilon,\mathrm{im}}^+$.
Of course if $z \in \mathcal{U}_{\varepsilon,\mathrm{im}}^+ \cap \mathcal{Z}(\mathcal{U}_\varepsilon^+)$ $[z, K_\lambda] = 0$, $[z, E_i] = 0$ in $\mathcal{U}_\varepsilon$ $\forall i \in I$ and more generally $[z, E] = 0$ in $\mathcal{U}_\varepsilon$ $\forall E \in \mathcal{U}_\varepsilon^+$.
Remark also that $\forall i \in I_0$ $F_i = -K_i T_{\omega_i}^{-1}(E_{\delta-\alpha_i})$. Then $\forall i \in I_0$

$$[z, F_i] = [z, -K_i T_{\omega_i}^{-1}(E_{\delta-\alpha_i})] = -K_i T_{\omega_i}^{-1}([z, E_{\delta-\alpha_i}]) = 0.$$

The claim follows applying lemma 3.4.1. □

Definition 3.
Following remark 3.2.9 (for the definition of J and f, see the notations used in the proof of theorem 3.5.1), I denote by $I_{\mathcal{Z}}$ the ideal of $\mathcal{U}_\varepsilon$ generated by

$$\{x_\alpha^{f(\alpha)}, \Omega(x_\alpha)^{f(\alpha)} | \alpha \in J\}.$$

Remark 4.
$\mathcal{U}_\varepsilon / I_{\mathcal{Z}}$ is a $\mathbb{C}$-algebra with basis

$$\{\Omega(x)(\underline{\gamma}) K_\lambda x(\underline{\tilde{\gamma}}) | \lambda \in Q \text{ and } \mu_{\underline{\gamma}}^\alpha, \mu_{\underline{\tilde{\gamma}}}^\alpha < f(\alpha)\ \forall \alpha \in J\}.$$

Proof: The claim is obvious because $x_\alpha^{f(\alpha)}, \Omega(x)_\alpha^{f(\alpha)} \in \mathcal{Z}(\mathcal{U}_\varepsilon)$ $\forall \alpha \in J$ and

$$\{\Omega(x)(\underline{\gamma}) K_\lambda x(\underline{\tilde{\gamma}}) | \lambda \in Q, \underline{\gamma}, \underline{\tilde{\gamma}} \in \cup_{\eta \geq 0} \mathrm{Par}(\eta)\}$$

is a basis of $\mathcal{U}_\varepsilon$. □

Proposition 5.

$$\mathcal{Z}(\mathcal{U}_\varepsilon/I_\mathcal{Z}) = \bigoplus_{\substack{\lambda\in Q: \\ l|(\lambda|\alpha_i)\ \forall i\in I}} \mathbb{C}K_\lambda .$$

Proof: Of course $\bigoplus_{\substack{\lambda\in Q: \\ l|(\lambda|\alpha_i)\ \forall i\in I}} \mathbb{C}K_\lambda \subseteq \mathcal{Z}(\mathcal{U}_\varepsilon/I_\mathcal{Z})$.
Notice now that $I_\mathcal{Z} = \oplus_{\alpha\in Q}(I_\mathcal{Z}\cap\mathcal{U}_{\varepsilon,\alpha})$; then $\mathcal{U}_\varepsilon/I_\mathcal{Z}$ inherits from $\mathcal{U}_\varepsilon$ a Q-gradation, that is

$$\mathcal{U}_\varepsilon/I_\mathcal{Z} = \bigoplus_{\alpha\in Q}(\mathcal{U}_\varepsilon/I_\mathcal{Z})_\alpha = \bigoplus_{\alpha\in Q}\mathcal{U}_{\varepsilon,\alpha}/(I_\mathcal{Z}\cap\mathcal{U}_{\varepsilon,\alpha}).$$

Then also its center is Q-graded because if $z_\alpha\in\mathcal{U}_{\varepsilon,\alpha}$ $(\alpha\in Q)$ and $x_\beta\in\mathcal{U}_{\varepsilon,\beta}$,

$$0 = \Big[\sum_{\alpha\in Q} z_\alpha, x_\beta\Big] = \sum_{\alpha\in Q}[z_\alpha, x_\beta] \quad\Rightarrow$$

$$[z_\alpha, x_\beta] = 0 \quad \forall\alpha\in Q.$$

Let

$$z = \sum_{\lambda,\underline{\gamma},\underline{\tilde\gamma}} a_{\lambda,\underline{\gamma},\underline{\tilde\gamma}} F(\underline{\gamma})K_\lambda E(\underline{\tilde\gamma}) \in (\mathcal{Z}(\mathcal{U}_\varepsilon)/I_\mathcal{Z})_\alpha$$

where I can suppose $\mu^\beta_{\underline{\gamma}}, \mu^\beta_{\underline{\tilde\gamma}} < f(\beta)$ $\forall\beta\in J$ because of the definition of $I_\mathcal{Z}$ (see definition 3.5.3).
Then $\forall\beta\in\tilde{R}_+$

$$0 = [z, E_\beta] = \sum_{\lambda,\underline{\gamma},\underline{\tilde\gamma}} a_{\lambda,\underline{\gamma},\underline{\tilde\gamma}}[F(\underline{\gamma})K_\lambda E(\underline{\tilde\gamma}), E_\beta] =$$

$$= \sum_{\lambda,\underline{\gamma},\underline{\tilde\gamma}} a_{\lambda,\underline{\gamma},\underline{\tilde\gamma}}F(\underline{\gamma})[K_\lambda E(\underline{\tilde\gamma}), E_\beta] + \sum_{\lambda,\underline{\gamma},\underline{\tilde\gamma}} a_{\lambda,\underline{\gamma},\underline{\tilde\gamma}}[F(\underline{\gamma}), E_\beta]K_\lambda E(\underline{\tilde\gamma}).$$

Remark that if $a_{\lambda,\underline{\gamma},\underline{\tilde\gamma}} \neq 0$ and $\underline{\gamma}\in\mathrm{Par}(\eta)$, $\underline{\tilde\gamma}\in\mathrm{Par}(\tilde\eta)$, then $\tilde\eta = \eta+\alpha$; then η is maximal in

$$\{\beta\in Q|\exists a_{\lambda,\underline{\gamma},\underline{\tilde\gamma}}\neq 0 \text{ with } \underline{\gamma}\in\mathrm{Par}(\beta)\}$$

if and only if $\tilde\eta$ is maximal in

$$\{\beta\in Q|\exists a_{\lambda,\underline{\gamma},\underline{\tilde\gamma}}\neq 0 \text{ with } \underline{\tilde\gamma}\in\mathrm{Par}(\beta)\}.$$

Remark also that if $\underline{\gamma}\in\mathrm{Par}(\eta)$

$$[F(\underline{\gamma}), E_\beta]K_\lambda E(\underline{\tilde\gamma}) \in \oplus_{\xi<\eta}\mathcal{U}^-_{\varepsilon,-\xi}\mathcal{U}^0_\varepsilon\mathcal{U}^+_\varepsilon .$$

In particular, if η_0 is maximal in $\{\eta\in Q|\exists a_{\lambda,\underline{\gamma},\underline{\tilde{\gamma}}}\neq 0 \text{ with } \underline{\gamma}\in\mathrm{Par}(\eta)\}$,

$$\sum_{\substack{\lambda,\underline{\gamma},\underline{\tilde{\gamma}}:\\ \underline{\gamma}\in\mathrm{Par}(\eta_0)}} a_{\lambda,\underline{\gamma},\underline{\tilde{\gamma}}}F(\underline{\gamma})[K_\lambda E(\underline{\tilde{\gamma}}),E_\beta]=0 \quad \forall\beta\in\tilde{R}_+.$$

Since $[K_\lambda E(\underline{\tilde{\gamma}}),E_\beta]\in\mathcal{U}_\varepsilon^{\geq 0}$, this implies that $\forall\underline{\gamma}\in\mathrm{Par}(\eta_0)$, $\forall\beta\in\tilde{R}_+$,

$$\sum_{\lambda,\underline{\tilde{\gamma}}} a_{\lambda,\underline{\gamma},\underline{\tilde{\gamma}}}[K_\lambda E(\underline{\tilde{\gamma}}),E_\beta]=0.$$

But $[K_\lambda E(\underline{\tilde{\gamma}}),E_\beta]=K_\lambda(E(\underline{\tilde{\gamma}})E_\beta-\varepsilon^{-(\lambda|\beta)}E_\beta E(\underline{\tilde{\gamma}}))$; then $\forall\beta\in\tilde{R}_+$, $\forall\underline{\gamma}\in\mathrm{Par}(\eta_0)$ and $\forall\lambda\in Q$,

$$\sum_{\underline{\tilde{\gamma}}} a_{\lambda,\underline{\gamma},\underline{\tilde{\gamma}}}(E(\underline{\tilde{\gamma}})E_\beta-\varepsilon^{-(\lambda|\beta)}E_\beta E(\underline{\tilde{\gamma}}))=0.$$

Now Levendorskii-Soibelman formula (see Beck [2]) says that

$$E(\underline{\tilde{\gamma}})E_\beta=aE(\vartheta_\beta(\underline{\tilde{\gamma}}))+\sum_{\underline{\gamma}'\succ\vartheta_\beta(\underline{\tilde{\gamma}})} a_{\underline{\gamma}'}E(\underline{\gamma}')$$

and

$$E_\beta E(\underline{\tilde{\gamma}})=\varepsilon^{r(\underline{\tilde{\gamma}};\beta)}aE(\underline{\vartheta_\beta(\tilde{\gamma})})+\sum_{\underline{\gamma}'\succ\vartheta_\beta(\underline{\tilde{\gamma}})} b_{\underline{\gamma}'}E(\underline{\gamma}')$$

where

$$r(\underline{\tilde{\gamma}};\beta)=\sum_{\tilde{\gamma}_i\succ\beta}(p(\tilde{\gamma}_i)|p(\beta))-\sum_{\tilde{\gamma}_i\prec\beta}(p(\tilde{\gamma}_i)|p(\beta)).$$

Then if $\underline{\tilde{\gamma}}\in\mathrm{Par}(\eta_0+\alpha)$ is minimal in $\{\underline{\tilde{\gamma}}'\in\mathrm{Par}(\eta_0+\alpha)|a_{\lambda,\underline{\gamma},\underline{\tilde{\gamma}}'}\neq 0\}$ also $\vartheta_\beta(\underline{\tilde{\gamma}})$ is minimal in $\{\vartheta_\beta(\underline{\tilde{\gamma}}')\in\mathrm{Par}(\eta_0+\alpha+\beta)|a_{\lambda,\underline{\gamma},\underline{\tilde{\gamma}}'}\neq 0\}$, so that one has that $\forall\beta\in\tilde{R}_+$ $\varepsilon^{r(\underline{\tilde{\gamma}};\beta)}=\varepsilon^{(\lambda|\beta)}$, that is $l|\big(r(\underline{\tilde{\gamma}};\beta)-(\lambda|\beta)\big)$.
If not all of the $\tilde{\gamma}_i$ are imaginary, there exists i such that $\tilde{\gamma}_i\in R_+^{\mathrm{re}}$ and either
i) $\tilde{\gamma}_j\succ\tilde{\gamma}_i\Rightarrow\tilde{\gamma}_j\succeq\tilde{\alpha}$ for some imaginary root $\tilde{\alpha}$, or
ii) $\tilde{\gamma}_j\prec\tilde{\gamma}_i\Rightarrow\tilde{\gamma}_j\preceq\tilde{\alpha}$ for some imaginary root $\tilde{\alpha}$.
Then

$$0=(\lambda|\tilde{\gamma}_i)-(\lambda|\tilde{\gamma}_i\pm\delta)\equiv_{(l)}r(\underline{\tilde{\gamma}};\tilde{\gamma}_i)-r(\underline{\tilde{\gamma}};\tilde{\gamma}_i\pm\delta)=$$

$$=\sum_{\tilde{\gamma}_j\succ\tilde{\gamma}_i}(p(\tilde{\gamma}_j)|\tilde{\gamma}_i)-\sum_{\tilde{\gamma}_j\prec\tilde{\gamma}_i}(p(\tilde{\gamma}_j)|\tilde{\gamma}_i)+$$

$$-\sum_{\tilde{\gamma}_j\succ\tilde{\gamma}_i\pm\delta}(p(\tilde{\gamma}_j)|\tilde{\gamma}_i\pm\delta)+\sum_{\tilde{\gamma}_j\prec\tilde{\gamma}_i\pm\delta}(p(\tilde{\gamma}_j)|\tilde{\gamma}_i\pm\delta)=$$

$$= \pm \sum_{\tilde{\gamma}_j = \tilde{\gamma}_i} (p(\tilde{\gamma}_j)|\tilde{\gamma}_i) = \pm\mu_{\underline{\tilde{\gamma}}}^{\tilde{\gamma}_i}(\tilde{\gamma}_i|\tilde{\gamma}_i),$$

from which $l|\mu_{\underline{\tilde{\gamma}}}^{\tilde{\gamma}_i}(\tilde{\gamma}_i|\tilde{\gamma}_i)$, that is $f(\tilde{\gamma}_i) = l_{\tilde{\gamma}_i}|\mu_{\underline{\tilde{\gamma}}}^{\tilde{\gamma}_i}$, which is impossible, because $1 \leq \mu_{\underline{\tilde{\gamma}}}^{\tilde{\gamma}_i} < f(\tilde{\gamma}_i)$.
Hence $\tilde{\gamma}_i \in R_+^{\mathrm{im}} \times I_0$ $\forall i$; from this it follows that if $\mathrm{Par}(\eta_0 + \alpha) \ni \underline{\tilde{\gamma}}' \succ \underline{\tilde{\gamma}}$ then $\tilde{\gamma}_i' \in R_+^{\mathrm{im}} \times I_0$ $\forall i$ and in particular

$$\forall \underline{\gamma} \in \mathrm{Par}(\eta_0), \quad \forall \lambda \in Q \quad \sum_{\underline{\tilde{\gamma}}} a_{\lambda,\underline{\gamma},\underline{\tilde{\gamma}}} E(\underline{\tilde{\gamma}}) \in \mathcal{U}_{\varepsilon,\mathrm{im}}^+;$$

on the other hand $r(\underline{\tilde{\gamma}};\beta) = 0$ $\forall \beta \in Q$ (because $\tilde{\gamma}_i \in \ker(\cdot|\cdot)$ $\forall i$) so that

$$l|(\lambda|\beta) \quad \forall \beta \in Q \quad \text{and}$$

$$\sum_{\underline{\tilde{\gamma}}} a_{\lambda,\underline{\gamma},\underline{\tilde{\gamma}}} \big(E(\underline{\tilde{\gamma}})E_\beta - \varepsilon^{-(\lambda|\beta)} E_\beta E(\underline{\tilde{\gamma}})\big) = \Big[\sum_{\underline{\tilde{\gamma}}} a_{\lambda,\underline{\gamma},\underline{\tilde{\gamma}}} E(\underline{\tilde{\gamma}}), E_\beta\Big] = 0.$$

This facts, together with lemma 3.5.2, imply that $\sum_{\underline{\tilde{\gamma}}} a_{\lambda,\underline{\gamma},\underline{\tilde{\gamma}}} E(\underline{\tilde{\gamma}}) \in \mathcal{Z}(\mathcal{U}_\varepsilon) \cap \mathcal{U}_\varepsilon^+$ hence the choice of $z \notin I_{\mathcal{Z}}$ implies that if, $a_{\lambda,\underline{\gamma},\underline{\tilde{\gamma}}} \neq 0$, $\mu_{\underline{\tilde{\gamma}}}^{\beta} = 0$ $\forall \beta$ (because $f(\tilde{\gamma}_i) = 1$ $\forall i$), and, since $\underline{\tilde{\gamma}} \in \mathrm{Par}(\eta_0 + \alpha)$, $\eta_0 + \alpha = 0$.
Then $a_{\lambda,\underline{\gamma},\underline{\tilde{\gamma}}} \neq 0$ implies $\underline{\tilde{\gamma}} \in \mathrm{Par}(0)$ and $l|(\lambda|\alpha_i)$ $\forall i \in I$, so that

$$z = \sum_{\substack{\lambda \in Q : l|(\lambda|\alpha_i) \\ \underline{\gamma} \in \mathrm{Par}(-\alpha)}} a_{\lambda,\underline{\gamma},\emptyset} F(\underline{\gamma}) K_\lambda \in \mathcal{U}_{\varepsilon,\alpha}$$

with $\alpha \leq 0$.
But $\Omega(z) \in \mathcal{U}_{\varepsilon,-\alpha}$ is also central, hence both α and $-\alpha$ are in $-Q_+$, that is $\alpha = 0$ and $z = \sum_{\substack{\lambda \in Q: \\ l|(\lambda|\alpha_i)}} a_{\lambda,\emptyset,\emptyset} K_\lambda$, which is the claim. □

Theorem 6.

$$\mathcal{Z}(\mathcal{U}_\varepsilon) = \mathcal{Z}^- \otimes_{\mathbb{C}} \mathcal{Z}^0 \otimes_{\mathbb{C}} \mathcal{Z}^+,$$

where

$$\mathcal{Z}^- \doteq \mathcal{Z}(\mathcal{U}_\varepsilon) \cap \mathcal{U}_\varepsilon^-, \quad \mathcal{Z}^+ \doteq \Omega(\mathcal{Z}^-) = \mathcal{Z}(\mathcal{U}_\varepsilon) \cap \mathcal{U}_\varepsilon^+,$$

$$\mathcal{Z}^0 \doteq \mathcal{Z}(\mathcal{U}_\varepsilon) \cap \mathcal{U}_\varepsilon^0 = \oplus_{\substack{\lambda : \forall i \in I \\ l|(\lambda|\alpha_i)}} K_\lambda \mathbb{C},$$

that is, $\mathcal{Z}(\mathcal{U}_\varepsilon)$ is the algebra of polynomials

$$\mathbb{C}[E_\alpha^{l_\alpha}, E_{(l_i m\delta,i)}, E_{r\delta}, F_\alpha^{l_\alpha}, F_{(l_i m\delta,i)}, F_{r\delta}, K_\lambda] \big/ (K_\lambda K_\mu - K_{\lambda+\mu}, K_0 - 1)$$

where $\alpha\in R^{\mathrm{re}}_+$, $m,r>0$, $i\in I_0$ and $\bar{c}_1^{(r)}=0$ in $\mathcal{U}_\varepsilon$ and $l|(\lambda|\alpha_i)$ $\forall i\in I_0$.

Proof: It is enough to apply lemma 3.2.1 and proposition 3.5.5, setting:

$$\mathcal{V}=\tilde{\mathcal{V}}\doteqdot\mathcal{U}_\varepsilon,\quad I\doteqdot I_{\mathcal{Z}},$$

$$T'=S'\doteqdot\{x_\alpha,\Omega(x_\alpha)|\alpha\in\tilde{R}_+\},\quad J'\doteqdot\{x_\alpha,\Omega(x_\alpha)|\alpha\in J\},$$

$$T''=S''\doteqdot\{K_\lambda|\lambda\in Q\},\quad J''\doteqdot\{K_\lambda|l|(\lambda|\alpha_i)\ \forall i\in I\},$$

$$f'(x_\alpha)\doteqdot f(\alpha),\quad f'(\Omega(x_\alpha))\doteqdot f(\alpha)\ \ \forall\alpha\in J.$$

□

REFERENCES.

[1] Beck, J., Braid group action and quantum affine algebras, Commun. Math. Phys. **165** (1994), 555-568.

[2] Beck, J., Convex bases of PBW type for quantum affine algebras, Commun. Math. Phys. **165** (1994), 193-199.

[3] Beck, J., Kac, V.G., Finite dimensional representations of quantum affine algebras at roots of unity, to appear Commun. Math. Phys.

[4] Bourbaki, N., Groupes et algèbres de Lie 4, 5, 6, Hermann, Paris, 1968.

[5] Damiani I., A basis of type Poincaré-Birkhoff-Witt for the quantum algebra of $\widehat{sl}(2)$, J. Alg. **161** (1993), no. 2, 291-310.

[6] De Concini, C., Kac, V.G., Representations of quantum groups at roots of 1, Progr. in Math. **92** (1990), 471-506, Birkhäuser.

[7] Drinfeld, V.G., A new realization of Yangians and quantized affine algebras, Soviet Math. Dokl. **36** (1988), no. 2, 212-216.

[8] Drinfeld, V.G., Hopf algebras and quantum Yang-Baxter equation, Soviet Math. Dokl. **32** (1985), 254-258.

[9] Drinfeld, V.G., Quantum groups, Proc. ICM, Berkeley (1986), 798-820.

[10] Humphreys, J.E., Introduction to Lie Algebras and Representation Theory, Springer-Verlag, USA (1972).

[11] Jimbo, M., A q-difference analogue of $\mathcal{U}(\mathfrak{g})$ and the Yang-Baxter equation, Lett. Math. Phys. **10** (1985), 63-69.

[12] Kac, V.G., Infinite Dimensional Lie Algebras, Birkhäuser Boston, Inc., USA (1983).

[13] Kac, V.G., Kazhdan, D.A., Structure of representations with highest weight of infinite-dimensional Lie algebras, Adv. Math. **34** (1979), 97-108.

[14] Levendorskii, S.Z., Soibelman, Ya.S., Some applications of quantum Weyl group I, J. Geom. Phys. **7** (1990), 241-254.

[15] Lusztig, G., Finite-dimensional Hopf algebras arising from quantum groups, J. Amer. Math. Soc. **3** (1990), 257-296.

[16] Lusztig, G., Introduction to quantum groups, Birkhäuser Boston, USA (1993).

[17] Lusztig, G., Quantum deformations of certain simple modules over enveloping algebras, Adv. in Math. **70** (1988), 237-249.

[18] Lusztig, G., Quantum groups at roots of 1, Geom. Ded. **35** (1990), 89-113.

[19] Matsumoto, H., Générateurs et rélations des groupes de Weyl généralisés., C. R. Acad. Sci. Paris **T. 258 II** (1964), 3419-3422.

[20] Rosso, M., An Analogue of P.B.W. Theorem and the Universal R-Matrix for $U_h sl(N+1)$, Commun. Math. Phys. **124** (1989), 307.

[21] Rosso, M., Finite dimensional representations of the quantum analogue of the enveloping algebra of a complex simple Lie algebra, Commun. Math. Phys. **117** (1988), 581-593.

[22] Shapovalov, N.N., On a bilinear form on the universal enveloping algebra of a complex semisimple Lie algebra, Functional Anal. Appl. **6** (1972), 307-312.

[23] Tanisaki, T., Killing forms, Harish-Chandra isomorphisms and universal R-matrices for Quantum Algebras, Int. J. Mod. Phys. **7** (1992), 941-962.

Elenco delle Tesi di perfezionamento della Classe di Scienze
pubblicate dall'Anno Accademico 1992/93

HISAO FUJITA YASHIMA, *Equations de Navier–Stokes stochastiques non homogènes et applications*, 1992.

GIORGIO GAMBERINI, *The Minimal supersymmetric standard model and its phenomenological implications*, 1993.

CHIARA DE FABRITIIS, *Actions of Holomorphic Maps on Spaces of Holomorphic Functions*, 1994.

CARLO PETRONIO, *Standard Spines and 3-Manifolds*, 1995.

MARCO MANETTI, *Degenerations of Algebraic Surfaces and Applications to Moduli Problems*, 1995.

ILARIA DAMIANI, *Untwisted Affine Quantum Algebras: the Highest Coefficient of* $det H_\eta$ *and the Center at Odd Roots of 1*, 1995.

Pantograf s.n.c. - Via alla Stazione di Voltri, 2/A - Genova
Finito di stampare nel gennaio 1996